The Beginning of Time

by

Julie Phan Le
and
Hue Van Le

The contents of this work, including, but not limited to, the accuracy of events, people, and places depicted; opinions expressed; permission to use previously published materials included; and any advice given or actions advocated are solely the responsibility of the author, who assumes all liability for said work and indemnifies the publisher against any claims stemming from publication of the work.

All Rights Reserved
Copyright © 2014 by Julie Phan Le and Hue Van Le

No part of this book may be reproduced or transmitted, downloaded, distributed, reverse engineered, or stored in or introduced into any information storage and retrieval system, in any form or by any means, including photocopying and recording, whether electronic or mechanical, now known or hereinafter invented without permission in writing from the publisher.

Dorrance Publishing Co
701 Smithfield Street
Pittsburgh, PA 15222
Visit our website at www.dorrancebookstore.com

ISBN: 978-1-4809-0888-8
eISBN: 978-1-4809-0750-8

*The Birth of the Universe and
Its Expansion by the Bessel Function of Zero Order*

*The Nihility Gas Gave Birth to the One and
Only God Cao Dai/Jehovah by
Subatomic Theory*

*The Doctrine of Nihilism by
Physical and Supernatural Science*

by
Julie Phan Le and Hue Van Le

Contents

Foreword		vii
Chapter 1	The Birth of the Universe and the Birth of God Cao Dai/Jehovah	1
Chapter 2	The Mystery of the Creation	40
Chapter 3	Number Twelve Is God Cao Dai/Jehovah's Own Special Number	49
Chapter 4	The Constitution of Trinity in Unity	56
Chapter 5	The Doctrine of Nihilism	67
Chapter 6	The Will of God Cao Dai/Jehovah	77
Chapter 7	Intelligent Versus Genius and Allocation of Wealth	81

Foreword

After our retirement, my wife and I tried to do a lot of research about religion and science to bring the truth out in many areas covered by a substantial dark curtain of knowledge that we need to discover. In this book, we try to bring ideas that make people think and justify, but not take it for granted, and we do not try to persuade anybody to follow anything but to use their own judgment.

Talking about the 'Birth of the Universe' when the first thermonuclear fusion explosion took place in the virgin space like a giant fireball, which was very hot and unstable. The overall diameter of that fireball was approximately equal to four billion light years across, and all the products of the first thermonuclear explosion called 'plasma' were pushed outward, obeying the 'Bessel Function of Zero Order' to form the Universe. The power or energy required to push the extraordinary giant volume of the plasma said above from the center to the outmost of the primitive universe for seven billion years could be multi-trillion trillions times larger than the power of $E = mc^2$ written by Albert Einstein.

If the velocity of light C of 186,300 miles per second is the natural speed limit will make it more critical, even impossible, to build the Universe because the value of $E = mc^2$ is too small to be qualified for the work of building the Universe. The first thermonuclear explosion sized at four billion light years across could create an extraordinary emitting force or energy to power outward the substantial volume of plasma for nearly seven billion years by the energy equation $E = \frac{1}{2} MV^2$; by that reason V should be a hundred thousand times larger than the speed of light C so C is not a natural speed limit. In addition, during the birth of the Universe, all the plasma product particles, ions, subatomic particles, matter, anti-matter, elementary particles, etc...should have moved much faster than light particles or photons. (Please read the explanation of this matter in Chapter 1.)

The equation used to calculate the mass in motion written by Albert Einstein—$M_m = M_0/\sqrt{(1 - V^2/C^2)}$—is far less than adequate to be applicable because the speed of light C is not the natural speed limit and it does not have any factor related to the mass in motion and this equation only satisfies one and only one case of an electron forced to accelerate to 99 percent of speed of light C in a vacuum, so it is not a practical application for an object on Earth, in space, or outside of the Universe, as well. (Please see the new calculation of the 'Mass in Motion' in Chapter 1).

Another true fact that light does have the least mass, so-called non-zero mass particles; mass is the primary part of light that carries energy (heat). If light does not have mass, the heat will die out shortly after being emitted and mass will withhold energy (heat) for the long distance of trillions of miles away; otherwise, photons cannot carry heat. Light will bend when passed by an object influenced by gravity, and light particles or photons have mass, so it can be split in half like a bubble. The true practice of physics is that energy can be split indefinitely, so we use the energy-to-mass conversion. The phenomenon of a photon split in half captured by Irene Curie and her husband Frederic Joliot in the early 1930s, showing one photon converted into two particles. Extreme gravity can prove that any object carries least mass or non-zero mass particles. The mass of photon could be converted into energy (heat) by using a convex lens to concentrate a large amount of photons from strong sunlight; one piece of paper is placed at the focal point of the lens for a few minutes, the piece of paper will catch on fire, and the flames of that fire immediately emit light particles (photons). This example tells us that light particles have mass which, can be converted into energy (heat) and vice versa. We did find the value of the mass of a photon in Chapter 1.

Every several thousands of years, Almighty God Jehovah/Cao Đài comes to our Planet Earth No. 68 and creates a new religion to update his children with the change of knowledge and science advancement. My wife and I took a lot of time to study the oldest to the youngest religions in Viet Nam from Ba-La-Mon, Hinduism, Confucianism, Taoism, Buddhism, Christianity, and Cao Daiism to answer these following questions: Where are we from? What are we doing? Where are we going?

The latest religion is Cao Daiism, which was established in 1926 in Viet Nam and around the world; this religion is the combination of theologies of Confucianism, Buddhism, Taoism, and Christianity. We found that Cao Daiism has tremendous information about the Universe and Nature regarding physical science and supernatural science mystery to fill the emptiness of our souls. In Cao Daiism, God Cao Đài/Jehovah used 'Great Mystery' (Huyền Cơ) to communicate with top religious leaders. There is so much information that illustrates the birth of the Universe, the birth of God Cao Đài/Jehovah, how He built the Universe with the number twelve (12), and the creation of Nature including human beings. These bits of information are directly associated with natural science, physical science, and supernatural science. Since I am a nuclear engineer and my wife is a pharmacist by profession, we took our background of knowledge to study and found these answers for the three questions stated above. My wife got sick and passed away on June 19, 2012; she had always wanted to finish this book, and I promised her I would do so.

In this book, everyone can also find out the mass of a photon and the mass of a neutrino by calculations with respect to the law of physics; these values have been ignored by all scientists since the birth of physics.

We rewrite the new equation for more accuracy in calculating the mass in motion of any object or of any particle to replace the one written by Albert Einstein. We assure that our Universe does have a 'Beginning' and no 'End' with proof; this is totally different from most scientists, including Stephen Hawking and Albert Einstein (most scientists believe that our Universe has *no* Beginning and *no* End). As Albert Einstein confirmed, the speed of light is the natural speed limit; we prove that the speed of light is not a natural speed limit in order to open the new road for science and physics to advance. We make a clearly different description between the Kingdom of God Cao Đài/Jehovah and Paradise.

We only finished part of it because of the abundance of ideas and mysterious theologies; those require more time to research and study.

Julie Phan Le and Hue Van Le

Chapter 1

The Birth of the Universe and the Birth of God Jehovah/Cao Dai
Tien Ong Dai Bo Tat Ma Ha Tat

The following references are the key bits of information that helped us to research and study to find out the truth of the birth of the Universe and the truth of how God Jehovah/Cao Đài was born. We took this information and the primitive source of life, which are atomic and subatomic theories to discover the truth. The physical science and the supernatural science did have many factors and causes, which are compatible and need to be explored.

Reference 1:
Thánh Ngôn Hiệp Tuyển (Quyển Thứ Nhứt), page 31. Thursday, July 22, 1926. *Huyền Cơ do Ngọc Hoàng Thượng Đế giáo đạo Nam Phương*
"*Khí Hư Vô Sinh Có Một Thầy.*
Còn Mấy Đấng Thầy kể đó (Confucius, Buddha Siddhartha, Lao-Tse, Jesus Christ, etc.) *ai sanh?*
Ấy là Đạo. Các Con nên biết. Nếu không có Thầy, thì không có chi trong Càn-Khôn Thế-Giới nầy; mà nếu không có Hư-Vô Chi Khí, thì không có Thầy."

Reference 1 Translation:
The selection of "The Words of Almighty God" (Book 1), page 31; Thursday, July 22, 1926. The 'Great Mystery' was from Almighty God Cao Đài/Jehovah teaching religion for the Third Amnesty in the Orient.

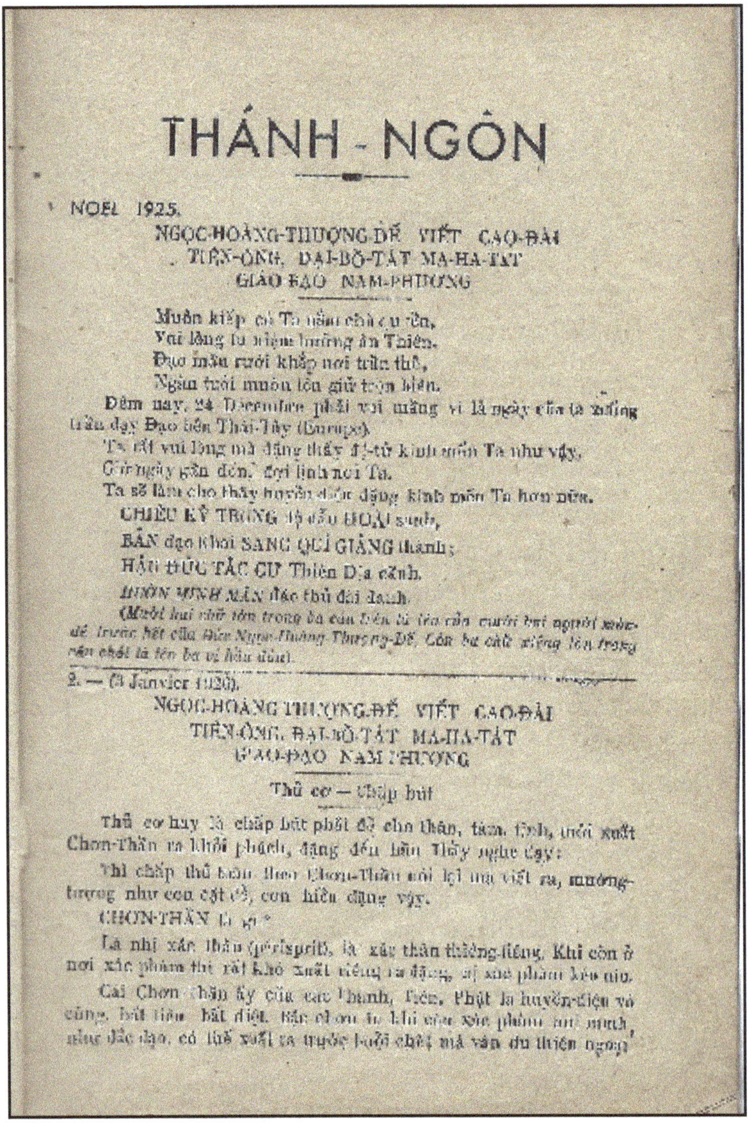

THÁNH - NGÔN

NOEL 1925.

NGỌC-HOÀNG-THƯỢNG-ĐẾ VIẾT CAO-ĐÀI
TIÊN-ÔNG, ĐẠI-BỒ-TÁT MA-HA-TÁT
GIÁO ĐẠO NAM-PHƯƠNG

Muôn kiếp có Ta nắm chủ quyền,
Vui lòng tu niệm hưởng ân Thiên.
Đạo mầu rưới khắp nơi trần thế,
Ngàn tuổi muôn tên giữ trọn biên.

Đêm nay, 24 Décembre phải vui mừng vì là ngày của Ta xuống trần dạy Đạo bên Thái-Tây (Europe).

Ta rất vui lòng mà đặng thấy đệ-tử kính mến Ta như vậy.

Gần ngày gần đến, đợi lịnh nơi Ta.

Ta sẽ làm cho thấy huyền diệu đặng kính mến Ta hơn nữa.

CHIÊU KỲ TRUNG độ dẫn HOÀI sanh,
BẢN đạo khai SANG QUÍ GIẢNG thanh;
HẬU ĐỨC TẮC CƯ Thiên Địa cảnh.
HÒN MINH MẪN đắc thù đài danh.

(Mười hai chữ tên trong ba câu trên là tên của mười hai người môn-đệ trước hết của Đức Ngọc-Hoàng Thượng-Đế. Còn ba chữ riêng lên trong câu chót là tên ba vị hầu đàn).

2. — (3 Janvier 1926).

NGỌC-HOÀNG-THƯỢNG-ĐẾ VIẾT CAO-ĐÀI
TIÊN-ÔNG, ĐẠI-BỒ-TÁT MA-HA-TÁT
GIÁO-ĐẠO NAM-PHƯƠNG

Thủ cơ — Chấp bút

Thủ cơ hay là chấp bút phải để cho thân, tâm, tịnh, mới xuất Chơn-Thần ra khỏi phàm, đặng đến hầu Thầy nghe dạy:

Thì chấp thủ hầu theo Chơn-Thần nói lại mà viết ra, mường-tượng như con đặt để, con hiểu đặng vậy.

CHƠN-THẦN là gì?

Là nhị xác thân (périsprit), là xác thân thiêng-liêng. Khi còn ở nơi xác phàm thì rất khó xuất riêng ra đặng, vì xác phàm kéo níu.

Cái Chơn-thần ấy của các Thánh, Tiên, Phật là huyền-diệu vô cùng, bất tiêu bất diệt. Bậc chơn tu khi còn xác phàm nơi mình, như đắc đạo, có thể xuất ra trước buổi chết mà vân du thiên ngoại

"The nihility gas gave birth to the one and only Thay (your master). Who gave birth to others such as Buddha, Siddhartha, Lao-Tse, Jesus Christ, etc.?" It was from Tao (way or religion). All my children need to understand. If Thầy (your master) does not exist, nothing will take place in this Universe; if there was no nihility gas, Thầy (your master) did not exist either." (The definition of the 'Nihility Gas' will be in Chapter 1.)

> -- 12 --
>
> Thập nhị khai Thiên là Thầy, Chúa cả Càn Khôn Thế Giới; nắm trọn Thập-Nhị Tuổi Thần vào tay. Số mười hai là số riêng của Thầy.
>
> ... Chưa phải hồi các con biết đặng tại sao vẽ Thánh Tượng "Con Mắt" mà thờ Thầy, song Thầy nói sơ lược cho hiểu chút đỉnh.
>
>   Nhãn thị chủ Tâm,
>   Lưỡng-quang Chủ-tể,
>   Quang thị Thần,
>   Thần thị Thiên
>   Thiên giã, ngã giã.
>
> Thần là khiếm-khuyết của cơ mầu nhiệm từ ngày Đạo bị bế. Lập « Tam Kỳ Phổ-Độ » nầy duy Thầy cho Thần hiệp «Tinh-Khí» đặng hiệp đủ «Tam Bửu» là cơ mầu nhiệm siêu-phàm nhập Thánh.
>
> Các con nhớ nói vì cớ nào thờ con mắt. Thầy cho chư đạo-hữu nghe...
>
> ... Phẩm vị Thần, Thánh, Tiên, Phật từ ngày bị bế Đạo, thì luật-lệ hỡi còn nguyên, luyện pháp chẳng đổi, song Thiên-Đình mỗi phen đánh tản «Thần» không cho hiệp cùng «Tinh Khí».
>
> Thầy đến đặng huờn-nguyên Chơn-Thần cho các con đắc đạo. Con hiểu «Thần cư tại Nhãn». Bố trí cho chư đạo-hữu các con hiểu rõ. Nguồn cội Tiên Phật do yếu-nhiệm là tại đó. Thầy khuyên con mỗi phen nói Đạo hằng nhớ đến danh Thầy.
>
> 13 Mars 1926.
>
> **NGỌC-HOÀNG THƯỢNG ĐẾ VIẾT CAO-ĐÀI
> GIÁO-ĐẠO NAM-PHƯƠNG**
>
> Thầy cho các con biết trước, đặng sau đừng trách rằng quyền-hành Thầy không đủ mà kềm thúc trọn cả Môn Đệ.

Reference 2:
   "My Research and Study on Electron Density and Temperature Measurements in a Low Pressure Magnetoplasma" (by Hue Le) at the Tennessee Technological University, August 1976.

Note: Cao Đài Tiên Ông Đại Bồ Tát Ma Ha Tát is the Holy name that God Jehovah proclaimed while teaching religion in the Orient for the Third Amnesty.

> — 81 —
>
> Như kê bên Phật Giáo hay tặng Nhiên-Đăng là Chưởng-giáo ; Nhiên-Đăng vốn sanh ra đời Hiên-Viên Huỳnh Đế,
>
> Người gọi Quan-Âm là Nữ Phật-Tông, mà Quan-Âm vốn là Từ-Hàng Đạo-Nhân hiện thân. Từ-Hàng lại sanh ra lúc Phong-Thần đời nhà Thương.
>
> Người gọi Thích-Ca Mâu-Ni là Phật Tổ, Thích-Ca vốn sanh ra đời nhà Châu.
>
> Người gọi Lão-Tử là Tiên Tổ-Giáo, thì Lão-Tử cũng sanh ra đời nhà Châu.
>
> Người gọi Jésus là Thánh Đạo Chưởng-giáo, thì Jésus lại sanh nhằm đời nhà Hớn.
>
> Thầy hỏi vậy chớ ai sanh ra các Đấng ấy ?
>
> Khi Hư-Vô sanh có một Thầy, Càn nối đúng Thầy kể đó ai sanh ? Ấy là Đạo. Các con nên biết.
>
> Nếu không Thầy, thì không có chi trong Càn-Khôn Thế Giới nầy ; mà nếu không có Hư-Vô Chi khí, thì không có Thầy.
>
> Dimanche 25 Juillet 1926. (16-6-B D.),
>
> **CAO ĐÀI**
>
> Cười . . . .
>
> T . . . ., con coi mặc Thiên-Phục có xấu gì đâu con ?
> Một ngày kia sắc-phục ấy, đời sẽ coi quí-trọng lắm.
> Con có ! con có biết những điền ấy bao giờ !
>
> Cười . . . .
>
> Mấy đứa Lễ-Sanh có đồ sắp-đặt sự nghiêm trang trong đàn cơ Thầy, chớ chẳng phải duy để đi lễ mà thôi, mỗi đại đạo phải dù mặt ; chúng nó phải ăn-mặc trang-hoàng hai đứa trước, hai đứa sau, xem sắp đặt sự thanh-tịnh. Thầy dạn các con như đàn nội chẳng nghiêm, Thầy không giáng, ba con nhớ nghe !
>
> Tr . . ., J . . ., K . , T . . , nghe :

Reference 3:

Thánh Ngôn Hiệp Tuyển (Quyển Thứ Nhứt), page 12.

"*Thập Nhị Khai Thiên là Thầy, Chúa cả Càn Khôn Thế-Giới; nắm trọn Thập Nhị Thời Thần vào tay. Số mười hai (12) là số riêng của Thầy.*"

Reference 3 Translation:

The collection of the Words of God Cao Đài Tiên Ông Đại Bồ Tat Ma Ha Tat/Jehovah (Book 1), page 12.

"I, Thầy (your master), am exactly the founder of the Twelve Signs of the Zodiac, the Master of the whole Universe, and the Commander of the Twelve Supernatural Timer Divines. The number twelve (12) is my own special number."

The Relationship between Physical Science and Supernatural Science

The purpose of this book, *The Beginning of Time*, is to prove the truth about how God Jehovah/Cao Đài Tiên Ông Đại Bồ Tát Ma Ha Tát was born in His own words. (The nihility gas gave birth to one and only Thầy, your master. If Thầy did not exist, nothing would take place in this universe; if there was no nihility gas, Thầy, your master, did not exist either.)

How the Universe was born by the plasma theory in conjunction with first thermonuclear fusion explosion, *Nổ ầm* or Big Bang (Big Boom).

The Birth of the Universe

Space always contains the virgin gas, hydrogen; the gas started to move and created space wind and gradually formed a huge whirlpool of powerful, turbulent gas. That extremely strong wind could have been created by the supernatural God who built the super universe, which is the 'Great Universe' of the universe systems, our Universe included.

A super dense gas kept swirling and sucked in more gas for trillions of years and the whirlpool kept growing larger, more than three billion light years across; the swirling compaction of gas generated an extreme gravity that pulled in a gigantic amount of the virgin space gas (hydrogen) at an extreme velocity within the whirlpool. The gas particles impinged on each other and generated an extremely high temperature, more than trillions of kelvin. Since the gravity of the whirlpool was so large (it was trillions of times larger than any current black hole's gravity), and it sucked in everything (hydrogen gas), caved in by its own extreme gravity, and then **popped** instantly;

the first thermonuclear fusion explosion took place like a gigantic fireball in the virgin space.

Scientists called it 'Big Bang,' the religion Cao Daiism called it 'Nổ ầm and the Buddhism called it *Um Ma Ni Ri*. Whatever people called the sound of that thermonuclear fusion explosion, it was the very unique finding of the birth of the Universe. The religion Cao Daiism did clarify that wherever the sound 'Nổ Ầm' reached, life was created; otherwise, nothing could happen. I also found this similar theology in Buddhism.

The atomic number Z of the hydrogen atom is equal to one (1); a single proton constitutes the nucleus of the hydrogen atom, which is a highly flammable gaseous element—it is the lightest of all gases and could be found abundantly in the Universe. At the moment that *Nổ Ầm*/ Big Bang, or Big Boom took place, the overall diameter of the fireball was larger than four billion light years across, and all electrons of hydrogen atoms were gaining higher energy and stripped off from their nucleus (protons); in other words, the gigantic fireball was exactly the first plasma medium of the first degree including an infinite number of first free electrons and first free protons and nothing else, since the nucleus of a hydrogen atom is a single proton. Right after this thermonuclear fusion explosion, the proton to proton reaction took place immediately.

At that moment, all protons could carry millions of trillions of kelvin, but the temperature of the first plasma medium was approximately hundreds of trillions of kelvin only. Protons were forced to smash into each other by high energy and velocity levels and fused; this reaction created anti-matter positrons, which are equivalent to a matter of electron subatomic particles. Elementary particles such as neutrinos, baryons, muons, etc., were also born by this process.

Randomly, protons were forced to impact each other and fusion took place; one of the two protons gave up its charge to its proton mate to become a neutron and formed a new product called deuterium. This nuclear reactivity continued and created many other things such as helium-3 (a new product of the combination of deuterium and a nearby proton). Shortly after helium-3 was created, then helium-3 and helium-3 were smashing each other, and normal helium-4 was created, and two protons were born. This process shows that the abundance of the virgin gas or hydrogen (proton) is the primary source of Creation for all of Nature and the whole Universe, and it also recreated its own existence; thereby, all Nature and the Universe will forever last and time will never end. The process of proton to proton activity continued to produce more new elements—positive ions, negative ions, matter electrons, anti-matter positrons—and more elementary particles—neutrinos, muons, baryons, leptons, photons, and more subatomic particles, etc.

The Definition of Nihliity Gas: Nihility means existent to be nonexistent and vice versa, similar to binary code of 0 and 1.

The matter electron and the anti-matter positron (Existent or Yes = 1) smashed into each other and annihilated each other, rendering void (Nonexistent or Zero = 0), then void (Zero = 0) to be replaced by matter electron and anti-matter positron (Existent or Yes = 1). This repetitive process similar to binary code 0 and 1, emitted two radiant light particles, or photons, and some energy was released ($e^-e^+ \rightarrow \gamma^*$ + energy, γ^* will gradually build the giant extreme light or *Thái Cực Đăng*). This is the most important 'Mystery Phenomena' to create nihility gas (*Khí Hư Vô*), which is related to the birth of God Jehovah/Cao Đài Tiên Ông Đại Bồ Tát Ma Ha Tát as specified by His own words said in the "Words of God Cao Đài/Jehovah" (see reference). Please see the illustration of the Birth of God Cao Đài/Jehovah later on this chapter. The thermonuclear fusion process stated above still takes place around the Universe at the present time, but not strongly enough to make the Universe expand.

Deuterium is an excellent fuel for thermonuclear fusion reactors and preferred by many scientists to produce clean energy. High technology has been used to pump high energy into deuterium fuel; deuterium atoms gain high energy and become ionized, and then are immediately forced through a strong magnetic field for high velocity to fuse. The product of this thermonuclear fusion is helium. Helium is non-radioactive and could be recycled in the factory. The nuclear radioactive reactivity only takes place within the nuclear reactor vessel, and no radioactive waste is exposed or handled in any way to harm the environment. The cost of this production of energy is very high because of the necessity of frequent replacement of the very expensive reactor vessel, due to rapid corrosion caused by high temperatures of hundreds of millions degrees F. Therefore, it is not an economical practice.

Deuterium could be found abundantly in sea water. During the first thermonuclear fusion explosion (Nổ Âm or Big Bang) in virgin space, ionization took place and produced a tremendous amount of ions, subatomic particles, plasma particles of different kinds, anti-matter, matter, radiant light particles, and halos; atomic mass split and drifted away from the nuclei of the virgin gas (hydrogen), in the form of phonons, gamma particles, or photons, etc.... These particles were pushing outward in all directions. The radial density distribution of all positive particles, negative particles, fusion products, subatomic particles, etc., are very unique and follow the Bessel function of zero order, similar to my research and study performed in 1976 at the Tennessee Technological University on 'The Electron Density and Temperature Measurements in a Low Pressure Magnetoplasma.'

With the great help of my professor, Dr. Carl Ventrice, the research of electron density and temperature measurements were made in an argon glow discharge at a pressure of one torr. The plasma was contained in a cylindrical chamber and ionized by a high DC voltage power supply. A triple electrostatic probe and a sensitive DC vacuum tube voltmeter were employed for the measurements. A uniform longitudinal magnetic field ranging from 0 to 650 gauss was used to create a magneto-plasma. The radial distribution of electron density was found to obey a Bessel function of zero order in every case. This was in agreement with plasma theory. The electron temperature was observed to be constant over the central portion of the plasma column, decreasing sharply as the wall was approached. This should be expected theoretically. The electron temperature on the tube axis was observed to increase monotonically with respect to the magnetic field.

The electron density on the tube axis was expected to increase as the magnetic field was increased, but it was observed to decrease in the magnetic field range between 100 and 275 gauss. This was possibly caused by the presence of instabilities in the plasma. It is important to note the uniqueness of this radial distribution, that is, its independence of all other gas parameters and of current. Since the Nổ Âm/Big Bang, or Big Boom took place right at the center of the giant fireball of extremely compact gas approximately three to four billion light years across, it did generate unimaginable heat to about a million trillion kelvin. The enormous amount of energy that is strong enough to power a substantial volume of the particles, ions, matter, anti-matter, and all other fusion products outward for nearly seven billion years, obeying a Bessel function of zero order. Approximately 99.999 percent of the whole Universe is under plasma status. The energy equation written by the scientist Albert Einstein is not powerful enough to power the gigantic volume of particles mentioned above for less than a few seconds.

$$E = mc^2 \qquad (1)$$
(C is the speed of light and equal to 186,300 miles per second.)

This energy equation is applicable in a small scale, such as making a nuclear bomb or generating enough heat to cook steam in a nuclear reactor to drive a turbine-generator to produce electricity only; this amount of energy is just enough to power a small volume of particles for several seconds, at most.

If this equation is applied to the birth of the Universe, the Universe could be equal to the size of the state of Alabama, USA, or smaller. During World War II, two nuclear bombs were dropped in two cities called Hiroshima and Nagasaki in Japan. Most of the products of thermonuclear bombs were traveling no more than a hundred miles from the centers of these two explosions. The power that was generated by the *Nổ Âm*/Big Bang pushed a substantial volume of the particles and plasma products for nearly seven billion years outward from the center of the first fireball to form the primitive universe sphere should require trillions of trillions of times larger than Albert Einstein's Energy Equation (1).

One should make a comparison between two identically powerful air rifles; one of them fires a solid BB shot and the other fires another BB shot which is empty inside.

The solid BB shot will travel a hundred times faster and ten times farther than the empty one. More power applied to similar case as mentioned above will result in greater speed and distance differences.

The first thermonuclear fusion explosion, *Nổ Âm* or Big Boom, caused by the extreme gravitational whirlpool of the gas (hydrogen) of the virgin space did create tremendous power to emit the substantial volume of subatomic and elementary particles, such as matter electrons, anti-matter positrons, neutrinos, neutrons, protons, baryons, leptons, etc., at the velocity of a hundred thousand times faster than the speed of light C.

These particles went through the gaps between atoms of any object or the virgin hydrogen cloud like a fast car traveling on an empty super highway, especially regarding heavy elementary particles.

Therefore, Albert Einstein's energy equation would no longer be applicable to the birth of the Universe and should be rewritten to match the power that gave birth to the primitive universe.

$$E = \tfrac{1}{2} M_0 V_m^2 \qquad (2)$$

Where:
- E is the energy of a moving object;
- M_0 is the mass of an object at rest; and
- V_m is the velocity of the moving object or of a particle, and it is a million times faster than the speed of light C equal to 300 thousand km per second.

This is the calculation of the mass in linear motion without speed of light.

As specified above, when the first thermonuclear fusion explosion took place, the temperature of the primitive universe was uniform inside and dropped sharply near the edge of the Universe. The radial distribution of all particles, ions, matter, anti-matter, and plasma products, etc., obeyed the Bessel function of zero order; in other words, the concentration of all particles were denser at the center of the Universe and gradually less dense as they approached the edge of the Universe. The mass of a moving object will proportionally change with respect to the square ratio of its moving velocity and its kick off velocity without limit:

$$M_m = M_0 + M_0 \Delta E \qquad (3)$$

Where:
M_m is the mass in motion;
M_0 is the mass at rest; and

$M_0 \Delta E$ is the net ratio of gain/loss of kinetic energy of the mass in motion at the select point m, compared with that of the kickoff point 0, and ΔE is the ratio change of the kinetic energy required to move the object from the kickoff point 0 to the point m selected on the line of motion. $M_0 \Delta E$ is the portion of mass gain/loss with respect to kinetic energy change ΔE.

Test:
$\Delta E = 0 \rightarrow$ Mass is at rest and equal to M_0, $\Delta E = 1 \rightarrow$ Mass in linear motion and equal to $2M_0$, $\Delta E \leq 1 \rightarrow$ Mass is losing its kinetic energy and moving at slower speed and proportionally decreased by ΔE and the mass is smaller than $2M_0$, $\Delta E \geq 1 \rightarrow$ Mass is gaining kinetic energy, moving faster and proportionally increased by the ratio ΔE and its mass is larger than $2M_0$.

More energy mass will move faster, less energy mass will move slower, and no energy mass will be at rest. Any parameter, such as gravity, that affects the mass system will directly affect its velocity and result in changing mass with it; to overcome the unwanted negative matter, energy and pushing force must be changed accordingly to compensate it.

The net change in mass will be proportional to the change of ΔE in the kinetic energy that is exerted upon it; it is also proportional to the ratio of its acceleration force, as well. Both of these applications will have the same result.

Since the kinetic energy of a moving object or of a particle is:

$$E = \tfrac{1}{2} M_0 V_m^2 \qquad \text{[same as Equation (2)]}$$

Where:
　E is the kinetic energy that powers an object or a particle;
　M_0 is the mass of an object or of a particle at rest;
　V_m is the velocity in motion of an object or of a particle at the point m; and
　V_0 is the kickoff velocity of an object or of a particle.

The net change of the kinetic energy that powers the object or a particle will be:

$$\Delta E = \frac{1/2 M_0 V_m^2}{1/2 M_0 V_0^2} \quad (4)$$

After simplifying:

$$\Delta E = V_m^2 / V_0^2$$

Equation (3) will be rewritten as follows:

$$M_m = M_0 + M_0(V_m^2/V_0^2)$$

$$M_m = M_0(1 + V_m^2/V_0^2) \quad (5)$$

This mass in motion calculation will work on Earth and in space or outside of the Universe, as well, because any parameter that affects the system will affect its velocity; in other words, the variation of the mass in motion will be solely affected by the square ratio of its traveling and kickoff velocities. There is no limit applicable on mass in motion and on its velocity, as well.

The variation of the mass in motion has nothing to do with the velocity of light C, and the velocity of light is not the natural speed limit either. As for the law of physics, the kinetic energy of an object or of a particle is written as in Equation (2):

$$E = 1/2 M_0 V_m^2 \quad (2)$$

This energy equation must be a trillion times greater than the energy equation written by Albert Einstein ($E = mc^2$).

Where:
    E is the energy required to power an object or a particle;
    M_0 is the mass of an object or of a particle;
    V_m is the velocity of an object or of a particle in motion.

V_m should be thousands of times larger than the velocity of light C equal to 186,300 miles per second, because the *Nổ Ầm*/Big Bang (the gigantic fireball or the first thermonuclear fusion explosion) mentioned above with the overall diameter approximately equal to four billion light-years across, could create an extremely gigantic power to push a substantial volume of thermonuclear product outward to support Universe expansion. Its power was infinite or at least multi-trillion trillion times larger than the power of the nuclear bomb dropped on the city of Hiroshima in Japan.

This requirement of an extraordinarily gigantic amount of energy shown in Equation (2) must be large enough to power outward a substantial volume of ions, subatomic particles, and plasma products created by the *Nổ Ầm*/Big Bang from the center to the outmost of the primitive universe for nearly seven billion years; it is obvious that V_m must be a hundred thousand times larger than the speed of light C to provide that energy demand. (Please see the example of the two BB shots fired by two identical powerful air rifles.) V_o is the essential factor of the kickoff velocity, which is associated with energy that moves the object and should always be considered as the primitive basic reference used to compare, to calculate, and to monitor the mass of the moving object or of a particle.

The mass in motion is calculated with respect to the force F that exerts upon an object or on a particle:

$$M_m = M_0 + M_0 \Delta F \qquad (6)$$

Where:
 M_m is mass in motion of an object or of a particle;
 M_0 is a mass at rest of an object or of a particle; and,
 ΔF is the ratio that changes the force that exerts upon an object or upon a particle.

$$\Delta F = \frac{F_m}{F_0} \qquad (7)$$

By Newton's law:
$$F = m\frac{d^2s}{dt^2} \quad (8)$$

Where s is the displacement.

$$F_m = M_0 a_m \quad \text{and} \quad F_0 = M_0 a_0 \quad (9)$$

Note: The metric system and inch system are applicable.

Where:
F_m is the force that is exerted upon an object at a selected point m, far away from the kickoff point;
F_0 is the kickoff force at the very first few seconds of t_0;

M_0 is the mass at rest;
d_m is the traveling distance of an object to the point m;
a_0 is the kickoff acceleration;
a_m is the acceleration of the object at the selected point m on the line of motion; and
a_m will be:

By Eq. (8) and (9):
$$a = \frac{d^2s}{dt^2} = \frac{(d_m - d_0)^2}{(t_m - t_0)^2} = \frac{d_m^2 - 2d_m d_0 + d_0^2}{t_m^2 - 2t_m t_0 + t_0^2} \quad (10)$$

Where $d_s = d_m - d_0$, and $t = t_m - t_0$ are traveling distance and traveling time respectively and kickoff distance d_0 and kickoff time t_0.

Several seconds, minutes, hours, months, or years after the very first couple of seconds of kickoff time t_0, the object will be far away from the starting point many miles away called traveling distant d_m.

Where:
d_0 is the kickoff distance measured in feet;
t_0 is the kickoff time, which is the first few seconds from kicking off;

d_m is the traveling distance measured in miles; and
t_m is the traveling time measured in minutes, hours, days, etc.

After several minutes, hours or days of traveling, t_0 and d_0 will be too small to be taken into consideration and Equation (10) will be simplified:

$$a_m = \frac{d_m^2}{t_m^2} = V_m^2 \qquad (11)$$

Right after the first few seconds of kickoff time, t_0, d_m, and t_m are too small to be taken into consideration. The kickoff acceleration would be:

$$a_0 = \frac{d_s^2}{dt^2} = \frac{(d_m - d_0)^2}{(t_m - t_0)^2} = \frac{d_m^2 - 2d_m d_0 + d_0^2}{t_m^2 - 2t_m t_0 + t_0^2} \qquad (12)$$

At the very first few seconds after the kickoff where d_m and t_m are too small to be taken into consideration.

Equation (12) will be simplified as follows:

$$a_0 = \frac{d_0^2}{t_0^2} = V_0^2$$

Equation (7) will be rewritten as follows:

$$\Delta F = \frac{F_m}{F_0} = \frac{M_0 a_m}{M_0 a_0} = \frac{M_0 V_m^2}{M_0 V_0^2} = \frac{V_m^2}{V_0^2} \qquad (13)$$

Equation (6) will be rewritten as follows without natural speed limit:

$$M_m = M_0 + M_0 \Delta F \quad (6) = M_0(1 + V_m^2/V_0^2) \quad (14)$$

The calculation of the mass in motion by using kinetic energy or by acceleration force (Newton's law) will give the same result—there is no limit applicable on the velocity or on the mass in motion of any object or of any particle at all. Equation (5) is the same as Equation (14).

If M_m is infinite mass, so V_m must be infinite velocity, and no natural speed limit is applicable. Equation ($E = \frac{1}{2} M_0 V_m^2$) must be a trillion times larger than ($E = mc^2$) to match the demand of power to build the Universe mentioned above.

General Note: The energy and the pushing force are considered positive as long as the mass system is moving in the wanted direction; otherwise, a negative situation will be taken into consideration.

Albert Einstein's equation for calculating the mass of an object in motion is:

$$M_m = M_0/\sqrt{(1 - V^2/C^2)} \quad (15)$$

Where:
M_m is the mass of an object in motion;
M_0 is the mass of the object at rest;
V is the velocity of an object in motion; and
C is the velocity of light equal to 186,300 miles per second.

In Equation (15), the traveling velocity of the object V is limited at the value of C, where C is much smaller than the velocity of elementary particles such as neutrino, baryon, lepton, etc.

This Equation (15) does not give the true value of M_m in many situations associated with kickoff and traveling speeds of the following examples.

For example, when launching a space shuttle of several tons or emitting an elementary particle such as a baryon at the kickoff velocity of 1 mile, 10 miles, 100 miles, and up to 1,000 miles per second, the mass of the space shuttle or of any particle calculated by Equation (15) does not vary that much or almost remains the same because the value C^2 is too large. C does not have any key factor associated with space shuttle mass systems on Earth or in the space either. C cannot be utilized as a key factor to compare, to calculate, or to monitor the traveling situation of any object at any moment in the kickoff or traveling at all; in other words, speed of light C is not a factor or essential element to be used in calculation of the mass of an object in motion. Albert Einstein really wanted to mention to everyone that the speed of light C is the natural speed limit and nothing else, but it is not true and Equation (15) is not applicable by any reason.

Equation (5) and Equation (14) will give all the true values of any object listed from elementary particles—neutrino, baryon, or lepton to space shuttle at any moment and at any location, regardless of any affected parameter. In addition, the velocity of light C is not the natural speed limit because the velocities of a heavy elementary particle like neutrino, baryon, or lepton, caused by the enormous emitting force created by Nổ Ầm/Big Bang would be billions of times faster than C. (Please read the example of two BB shots fired by two identically powerful air rifles as stated above.)

Light does have mass, which is too small that scientists cannot measure it. Photon (mass) could be split in halves (this phenomenon was captured by Irene Curie and her husband, Frederic Joliot, in the early 1930s). Using a convex lens to concentrate the sunlight at a hot noon and placing a piece of paper at the focal point of the convex lens, the light photons will bombard the surface of the piece of paper and, within minutes, the paper will catch on fire; this phenomenon certainly proves that the mass of photons were converted into energy (heat) and burns the paper. Then the flame immediately emits light photons right after that moment; in other words, energy (heat) was reconverted back to photon or mass, which agrees with the law of physics.

No other force could be found in the Universe greater than the force created by Nổ Ầm or Big Bang; most heavy subatomic particles and massy elementary particles were kicked out by extremely powerful forces and traveled billions of times faster than photons to form the universe sphere with respect to the Bessel function of zero order. Please see the Bessel function of zero order on next page.

Reference 2:
"The Research and Study on Electron Density and Temperature Measurement in a Low Pressure Magnetoplasma"

C is much slower than the velocity of heavy elementary particles like neutrino, baryon, and lepton, etc....and not the natural speed limit. Since a light particle or photon has the least mass and is invisible to scientist's measurement equipment, its momentum is also extremely small; therefore, the velocity of all photons maintains uniformity with their emitting force. As of Absolute Standards: F = ma (*F* is the force that pushes the unit mass m).

The mass of a photon m is extremely small as m = 9 x $10^{-9,999}$—or smaller—per assumption. Then F = ma = **a** x 9 x $10^{-9,999}$ will remain the same as F = 9 x $10^{-9,999}$ because the acceleration of the photon a = feet/sec^2 is very small, so the product of ma remains nearly the same, a so-called non-zero force applied for most of the non-zero mass particles, including photon. These non-zero mass particles are traveling at the same speed.

In other words, every photon receives the pushing force from any object that emits it with the magnitude approximately equal to its own weight; therefore, all photons always travel at the same speed of 186,300 miles per second. This is the known value to be used as reference, but it is not the natural speed limit. There are many other elementary particles as small as photon, but they are heavier than photon such as neutrinos, baryons, mesons, leptons, and muons, etc., when they receive the emitting force (F = ma), which is larger (mass comes into effect) and travel faster than photon similar to two example BB shots previously mentioned. This is an example to show the different speeds between heavy elementary particles and non-zero mass particles, including photons. The speed of light could be utilized as a known reference in common calculation, but not for natural speed limit. The mass of a moving object will proportionally change with respect to the square ratio of its moving velocity and its kickoff velocity without limit.

During the birth of the Universe, *Nổ Âm*/Big Bang, photons and elementary particles were emitted with the same power at the same time; photons had the least mass and traveled at a slower velocity than the heavier elementary (super subatomic) particles similar to

Dividing equation 2.143 by $2\pi r D_a^B dr$, one can get

$$\frac{d^2 n}{dr^2} + \frac{1}{r}\frac{dn}{dr} + \frac{\alpha u}{D_a^B} n = 0 \qquad (2.144)$$

The solution of equation 2.144 is the Bessel function of zero order given by equation 2.145

$$n(r) = n_0 J_0(r \cdot (u/D_a^B)^{1/2}) \qquad (2.145)$$

where n_0 is the particle concentration at $r = 0$. The equation 2.145 is the general solution to the differential equation 2.144 which indicates the radial distribution of both positive ion and electron densities $n(r)$. It is important to note the uniqueness of this distribution, that is, its independence of all other gas parameters and of current.

One may expect to determine the argument β of the zero-order Bessel function which is as follows:

$$n(r) = n_0 J_0(\beta r') \qquad (2.146)$$

Equation 2.146 is the equation 2.145 rewritten under the argument form of Bessel function where

$$\beta = \left[\frac{u}{D_a^B}\right]^{1/2} \qquad (2.147)$$

and

$$r' = \frac{r}{R} \qquad (2.148)$$

where R is the radius of positive column. One shall note that the argument β depends upon the interactions of plasma parameters, ionization coefficient, and ambipolar diffusion coefficient. In the case of no magnetic field applied to the plasma, D_a^B is replaced by D_a, and β is rewritten as follows:

$$\beta = [\frac{\alpha}{D_a}]^{1/2} \qquad (2.149)$$

Substituting equations 2.147 and 2.148 into equation 2.146, one can get

$$n(r) = n_o J_o(\beta \frac{r}{R}) \qquad (2.150)$$

One of the purposes of this paper is to investigate experimentally this theorical development. The results will be discussed in Chapters 4 and 5.

THE THEORY OF TRIPLE PROBE

Up to the present, a great number of studies have been made on various types of probe methods, and these are powerful tools for diagnosing plasma. The "single" and "double electrostatic probe," the "symmetric and asymmetric triple probe" are representative; each of these is useful for a slightly different purpose and has its own practical advantages.

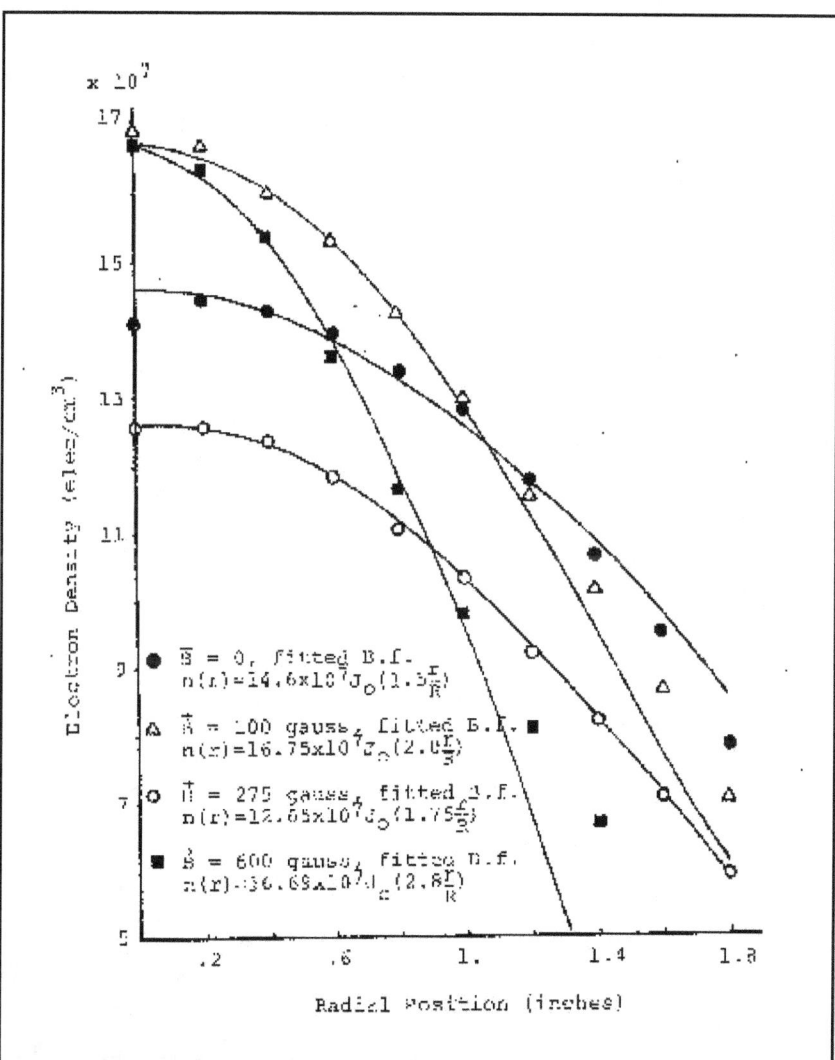

Fig. 4.1 Experimental Data Showing the Radial Distribution of the Electron Density for Different Magnetic Field Intensities

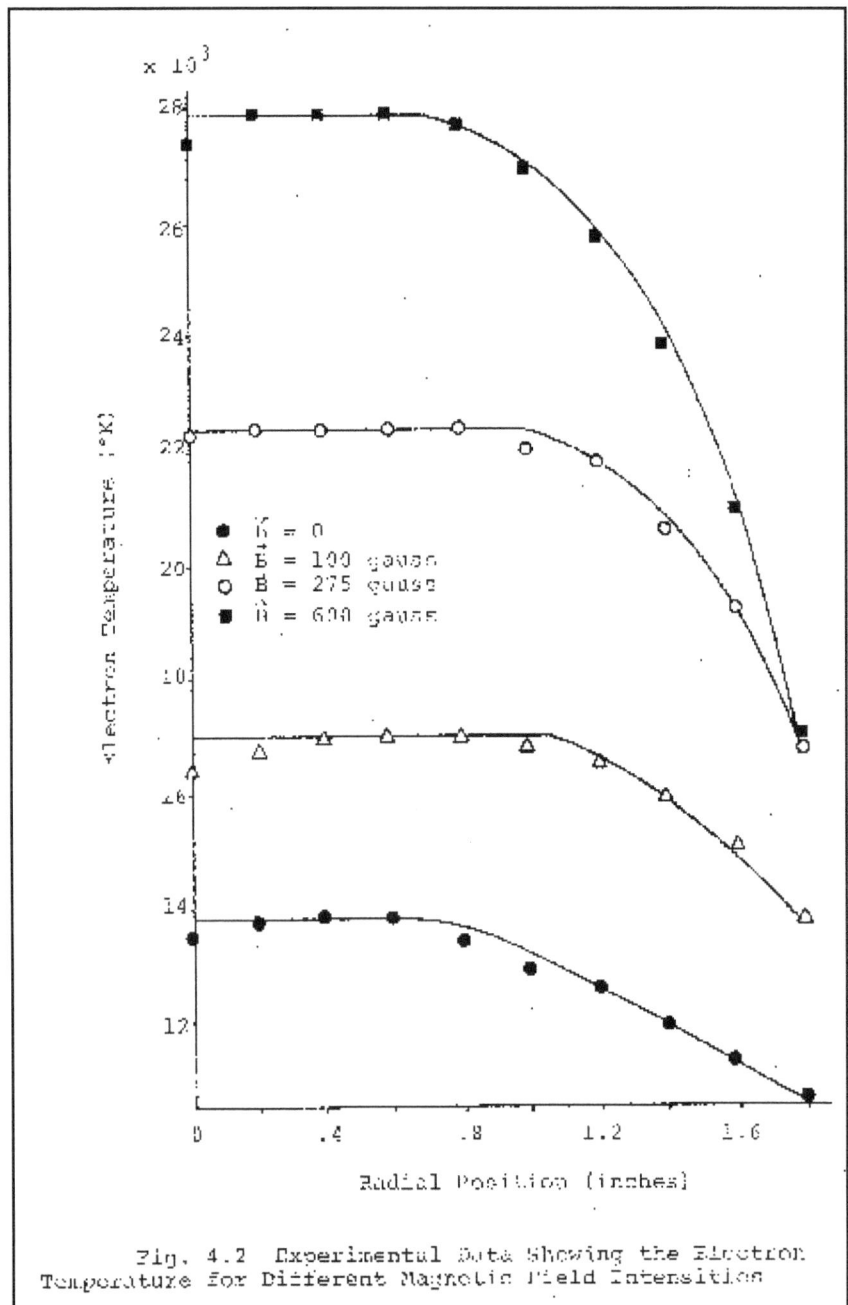

Fig. 4.2 Experimental Data Showing the Electron Temperature for Different Magnetic Field Intensities

the two BB shots fired by two identically powerful air rifles above. Light will bend when passed by a gravitational object. Light cannot escape the extremely gravitational 'Black Hole'; in other words, light does have mass.

Scientists can utilize extreme gravity to test anything that carries least mass. Heat cannot be transferred over a long distance without a mass to carry it; the least mass that photons carry is the primary part to hold energy (heat) for transferring trillions of trillions of miles away. One more example to show that photons have mass.

The Mass to Energy Conversion and Vice Versa

Using a high power convex lens to focus the bright sunlight at noon, place a piece of paper or cotton at the focal point where the highest concentration of light will be. The highest concentration of photons at the focal point will bombard the surface of said paper and will also smash into each other forcefully; within minutes, the paper will catch on fire. This phenomenon indicates that with the strong collision and bombardment force, the mass of concentrated photons will be converted into energy (heat) to burn the paper, then the fire (flame) will emit light (photons) immediately, and energy (heat) will be converted back into mass. These phenomena truly satisfy the law of physics, which is 'mass could be converted into energy and energy also could be converted back into mass.' This experiment alone can prove that light (photons) does carry mass. Without mass, heat can only travel a short distance and dies out. From a physics standpoint that energy could be split indefinitely and the energy continue to mass conversion. The phenomenon of a photon split in half was captured by Irene Curie and her husband Frederic Joliot in the early 1930s, indicating the conversion of one photon (mass of a photon) into two particles. This proves exactly that light carries mass for heat transfer; whenever light distributes heat (energy) or its mass is converted into energy, the light will cease immediately. The subatomic and elementary particles were heavy and blown away by enormous emitting forces created by the first thermonuclear fusion explosion sized approximately equal to four billion light-years across (Nổ Âm/Big Bang); they traveled at an extremely fast velocity a million times greater than the speed of light C. They passed through tiny gaps between the atoms of any object in their paths or untouched hydrogen atom clouds in the virgin space easily like a bullet. At the same time, light particles or photons were lighter and traveled at a slower speed like bubbles; they could not travel faster than massy elementary particles because of their tiny pushing non-zero force.

All positive ions, electrons, matter, anti-matter, plasma products, etc.…were traveling outward following a Bessel function of zero order for from one to six or even seven billion years (it depended on the size and weight of the particles) and started to degenerate (becoming less dense), at the same time they lost energy near the ground state.

One should find the higher density of all ionized gases, electrons, positrons, protons, elementary (super subatomic) particles, positive and negative ions at the center of the primitive gigantic fireball, and gradually less dense on the way outward to the outer space; this radial distribution obeys the Bessel function of zero order (see Reference 2 on Page 1). When the outward traveling slowed down and ceased, they formed the spherical seal of the primitive universe. The temperature at the border between the universal sphere and the outer space was approximately two to three kelvin.

There were large shockwaves, nuclear fusion reactivity, and super novas took place within the primitive universe, but not strong enough to support the universe expansion. Light photons eventually reached the universal seal after nearly seven billion years after the first thermonuclear fusion explosion, *Nổ Ầm*, or Big Bang. It is estimated that the temperature would drop from two to one kelvin at a distance of nearly ten light-years from the universal shell outward to the space.

The Mass of the Quantum of Light or Photon - by the Laws of Physics

Light is primarily the essential source of life for all Nature from tiny organisms, vegetation, and animals to all living creatures, including human beings. God Cao Đài/Jehovah pre-designed all living solar systems such that the quantum of light or photons could maintain a necessary amount of mass for a calculated distance (between star and living planet), such that mass to energy conversion could match exactly the necessary demand for all Nature in normal condition. If a photon travels only a short distance, when it bombards living nature such as tree leaves or human skin, the mass to energy conversion will be too large and the trees and human beings would be killed and no one could stay outdoors; however, if the planet is too far from the star, the quantum of light will not be able to deliver enough energy to protect life either. As said above, using a powerful convex lens to concentrate a large number of photons from the hot summer sunlight of a hot noon and placing a piece of paper at the focal point for a few minutes will cause the paper to catch fire.

Photons travel at high velocity and strongly bombard the surface of the paper; the total mass of the large amount of photons will be converted into a significant energy (heat) and burn the paper. Since the law of physics states that energy could be converted into mass and mass could be converted back into energy, the birth of the photon will own the same principle.

There are many ways to create photons; the simplest way to understand the process is that when an atom passes from one high energy state to another of lower energy, it delivers the extra energy to a single quantum of light or photon. If a photon does not have mass, it cannot carry energy and cannot travel, either, because it is an elementary particle like a neutrino, lepton, or baryon, etc. A photon is also an elementary particle, all of them borne from an atom; the photon is the one that has the least mass compared to other elementary particles and it is affected by the extreme gravity of a 'Black Hole.'

The Motion of a Photon:

Linear motion with respect to its linear momentum ($P = mc$).

All photons are borne mainly from the atoms; the starlight will tell us that the source of light is originated from hot atoms. Normally, high thermonuclear fusion reactivity takes place inside the core of stars at scorching temperatures of hundreds of millions of degrees F.; helium, proton, or deuterium, etc., are cooked at an extremely high temperature and are pushed violently outward to the surface of the stars.

These ions or hot atoms are also spinning at high angular momentum; when they reach the surface of the stars, all of them will pass from one high energy state to another of lower energy; the extra energy will be instantly converted into mass (quantum of light) and tossed out away from the cooler side of heated atoms at the velocity of 300 million meters per second like an athletic competitor tossing a dumbbell or an iron ball. Since a photon is an elementary particle and has least mass, it will spin and travel similar to a bullet fired by a rifle.

The Mass of the Quantum of Light or Photon

The angular velocity of the photon is too small and could not be measured easily in comparing it with its known linear velocity of 300 million meters per second; in other words, one should take the kinetic energy of the quantum of light (photon) into consideration:

$$E = \tfrac{1}{2} mv^2 \qquad (1a)$$

Where:
  E is the kinetic energy of a photon;
  m is the mass of a photon; and
  v is the velocity of a quantum of light or photon equal to C, which is 300 million meters per second.

E is a solely unique source of kinetic energy that a quantum of light could deliver and its angular momentum is too small to be considered. Once again, when an atom passes from one high energy state to another of lower energy, it will toss out a tiny mass of quantum of light.

Since a photon is an elementary particle, which has a tiny mass and travels at an extremely fast velocity of 300 million meters per second, when it bombards any atom on the surface of an object, it will pass through the electron cloud of that atom and violently bump its nucleus, instantly converting its mass into energy. The atom receives more and more photons in a similar way, resulting in gaining a higher energy level; when an atom is gaining a higher energy level, its electron cloud will be farther away from its nucleus and the outmost electron will be stripped off first, depending upon the receiving energy level. These photons do not disappear or are annihilated they're just converted from mass into an energy state. Modern physics normally specifies this phenomenon as disappearance and annihilation; it is not a close observation regarding the tiny mass of a photon.

Max Planck set the law of physics and the foundation for the advent of the quantum theory in the early twentieth century regarding unit energy of a quantum of light as:

$$E = h\nu \qquad (2a)$$

Where:
E is the energy that a unit of quantum of light could deliver;
h is Planck's constant and equal to ($= 6.626 \times 10^{-34}$ joule-second); and
ν is the frequency of the quantum of light or photon in hertz.

In 1979 the International System of Units (SI) defined the base unit candela as the luminous intensity, in a given direction, of a source that emits radiation of frequency of 540×10^{12} hertz. (This is the close value of ν to be chosen.)

Both Equations (1a) and (2a) represent the unit value of energy of the quantum of light or Equation (1a) equals Equation (2a).

$$h\nu = \tfrac{1}{2} m v^2 \qquad (3a)$$

$$m = \frac{2 h\nu}{v^2} \qquad (4a)$$

$$m = \frac{2 \times 6.626 \times 10^{-34} \times 540 \times 10^{12}}{(300{,}000{,}000)^2}$$

$$m = 795 \times 10^{-38} \text{ kg}$$

The mass of a quantum of light or photon is approximately equal to:

$$m = 7.95 \times 10^{-33} \text{ g}$$

The approximation of the mass of a photon is roughly several million times lighter than the mass of an electron. This is the one and only way to approach the mass of a photon and no science equipment could do so until the neutrino could replace the electron in the electronic world.

The normal practice of physics was that energy could be divided indefinitely; likewise, the energy to mass conversion would surely get the same conclusion.

The Approximate Calculation of the Mass of a Neutrino

All particles of the family of lepton have mass such as the electron and muon except the neutrino; this is not a good scientific conclusion because all of them are subatomic particles or elementary particles similar to a photon. Later in 2011, the European Organization for Nuclear Research found that a neutrino travels faster than the speed of light. The velocity of neutrino (V_n) is approximately twenty-five percent faster than the speed of light C equal to 300 million meters per second or V_n = 300 million meters per second x 1.25 = 375 million meters per second. This is an approximation only because the speed of the neutrino is not yet officially confirmed. Once again, the two identical powerful BB rifles fire two BB shots; one is solid and the other one is empty inside. The solid BB shot is traveling faster and farther than the empty one. All particles of the lepton family are borne the same way, but different in mass. The mother atoms are tossing them out by the same emitting power; however, they travel at different velocities for only one reason—the difference in mass. The neutrino could be compared to the solid BB shot, per se, and the Photon is similar to the empty BB shot; in other words, two particles were pushed by the same powerful force, but the heavier one will gain higher momentum and travel faster than the lighter one (mass comes into effect on momentum and/or velocity).

The mass of the neutrino could be calculated as follows:

$$V_n = 1.25\ C \quad (1b)$$

Atoms are tossing out all elementary particles with the same power; however, the differences between momentums and/or velocities of these subatomic particles will solely depend on the individual particle mass. (Please see to two BB shots mentioned above.) In other words, mass and velocity are relatively proportional to each other regarding uniform power applied on them.

$$V_n = \frac{m_n}{m_p}\ C \quad (2b)$$

Where:
V_n = Velocity of a Neutrino
C = Speed of Light or Photon
m_n = Mass of a Neutrino
m_p = Mass of a Photon Equal to 7.95×10^{-33} g

$$m_n = \frac{V_n \times m_p}{C} \quad (3b)$$

$$= \frac{1.25 \times C \times m_p}{C}$$

$$m_n = 1.25 \times m_p = 1.25 \times 7.95 \times 10^{-33}\ g$$

$$m_n = 9.94 \times 10^{-33}\ g$$

The mass of a neutrino is slightly heavier than the mass of a photon.
According to the words of God Cao Đài/Jehovah, there are 3,072 solar systems in the Universe; all their distances from every star to every living planet had been calculated by God Cao Đài/Jehovah in such a way that photons could deliver the right amount of energy to support life and maintain the prosperity of His children and Nature at all times.

The Twelve Supernatural Timer Divines always have absolute power to control and position all of the constellations of the zodiac, all stars, galaxies, and planets, etc., in the right orbit and location to maintain exact time and good weather to protect all children of God Cao Đài/Jehovah from havoc or universal catastrophe caused by divergent movement of the twelve constellations of the zodiac, the sun, the moon, the principle planets, etc.... (See Reference 3.) One should remember that everything from a tiny particle of dust to the giant galaxy or constellations of the zodiac and all things in the Universe were created by God Cao Đài/Jehovah.

The Birth of God
Cao Đài Tiên Ông Đại Bồ Tát Ma Ha Tát/Jehovah

Note: Cao Đài Tiên Ông Đại Bồ Tát Ma Ha Tát is the Holy name that God Jehovah proclaimed during teaching religion in the Orient for the Third Amnesty.

Reference 1:

The selection of the Words of Almighty God (Book 1), page 31; Thursday, July 22, 1926. The 'Great Mystery' was from Almighty God Cao Đài/Jehovah Teaching Religion for the Third Amnesty in the Orient. "The Nihility Gas gave birth to one and only Thay (your master). Who gave birth to others such as Confucius, Buddha (Siddhartha), Lao-Tse, Jesus Christ, etc...? It was from Tao (way or religion). All my children need to understand. If Thầy (your master) does not exist, thereby nothing will take place in this Universe; if there was no Nihility Gas thereby Thầy (your master) did not exist either."

Reference 1:

Definition of Nihility Gas: Something exists and then becomes non-existent and vice versa; these repetitive phenomena could be considered as binary code zero and one or one and zero (1 and 0) as stated above.

Definition by Nuclear Physics: (e^+ e^- → γ^* + energy) these radiant light particles, γ^*, actually built the extreme light, which is God Cao Đài/Jehovah's Supreme Spirit.

> — 31 —
>
> Như kể bên Phật Giáo hay tặng Nhiên-Đăng là Chưởng giáo ; Nhiên-Đăng vốn sanh ra đời Hiên-Viên Huỳnh Đế.
>
> Người gọi Quan-Âm là Nữ Phật-Tông, mà Quan-Âm vốn là Từ-Hàng Đạo-Nhơn hiến thân. Từ-Hàng lại sanh ra lúc Phong-Thần đời nhà Thương.
>
> Người gọi Thích-Ca Mâu-Ni là Phật Tổ, Thích-Ca vốn sanh ra đời nhà Châu.
>
> Người gọi Lão-Tử là Tiên Tả-Giáo, thì Lão-Tử cũng sanh ra đời nhà Châu.
>
> Người gọi Jésus là Thánh Đạo Chưởng-giáo, thì Jésus lại sanh nhằm đời nhà Hớn.
>
> Thầy hỏi vậy chớ ai sanh ra các Đấng ấy ?
>
> Khí Hư-Vô sanh có một Thầy, Càn nấy đúng Thầy kể đó ai sanh ? Ấy là Đạo. Các con nên biết.
>
> Nếu không Thầy, thì không có chi trong Càn-Khôn Thế Giới nầy ; mà nếu không có Hư-Vô Chi-Khí, thì không có Thầy.
>
> Dimanche 25 Juillet 1926. (16-6-B D.).
>
> **CAO ĐÀI**
>
> Cười . . . .
>
> T . . ., con coi mặc Thiên-Phục có xấu gì đâu con ?
> Một ngày kia sắc-phục ấy, đời sẽ coi quí-trọng lắm. Con đã l con có biết những điều ấy hay gi ?
>
> Cười . . . .
>
> Mấy đứa Lễ-Sanh có dỗ sắp-đặt sự nghiêm trang trong đàn cầu Thầy, chớ chẳng phải dạy đủ đi lễ mà thôi, mỗi đại đàn phải dủ mặt ; chúng nó phô ôn-mặc trang-hoàng hai đứa trước, hai đứa sau, xem sắp đặt sự thanh-tịnh. Thầy dắt các con như dắt nội chẳng nghiêm. Thầy không giảng, ba con nhớ nghe !
>
> Tr . ., J . . . K., T . . . nghe :

The matter electron and the anti-matter positron (Things Exist or Yes = 1) smashed into each other and annihilated each other or rendering each other void (Nonexistent or Zero = 0) and then electrons and positrons take place again right at the same voided place (Voided to be Replaced or Existence or Yes = 1).

Then they, again, smashed into each other and annihilated each other or voided (Voided or Zero = 0) each other. The alternative of Zero = 0 and One = 1 could be considered as the 'Nihility of Nature'; these phenomena gave birth to the doctrine of nihilism or God Cao

Dai/Jehovah went back to His primitive birth phenomenon and created this doctrine of nihilism. This process emitted two radiant light particles or photons and some energy was released; this is the most important mystery phenomena to create nihility gas (*Khí Hư Vô*), which is related to the birth of God Jehovah/Cao Đài Tiên Ông Đại Bồ Tát Ma Ha Tát as specified by His own words said in the Words of God Cao Đài/Jehovah. (See References 1 and 3.)

The Birth of God Cao Dai/Jehovah

After nearly one hundred million years the center of the primitive universe started to clear up and cool off; a large portion of nihility gas, dark matter, nuclei, matter (electrons), anti-matter (positrons), etc., remained.

Dark Matter: Dark matter is a tiny piece of mass split from proton to proton's violent impingement during the first thermonuclear fusion explosion. It does not have any physical property like an atom or a subatomic particle, it is just a blank piece of mass split away from its mother proton.

At first it carries an extremely high temperature of a million trillion degrees F., and whenever it gives up energy (heat) in space, it simply returns as a piece of mass and is invisible to a magnetic field, since it is a piece of mass and always affected by gravity.

All of a sudden, the nihility gas (the formation of matter or electrons, and anti-matter or positrons) popped up one giant extreme light (*Thái Cực Đăng*) shaped like a beautiful lotus right at the center of the Universe.

Nihility gas was the formation of anti-matter (positrons) and matter (electrons), and they were found abundantly in the Universe after the *Nổ Ầm* or Big Bang. Anti-matter and matter were smashing and annihilating each other and finally rendering each other void and emitted radiant light particles or photons.

God Cao Đài Tiên Ông Đại Bồ Tát Ma Ha Tát/Jehovah went back to the primitive source of His life and found the root of the nihility gas; He then created the doctrine of nihilism from here for religions Cao-Daiism and Buddhism. This highly spiritual theology of nihilism will be illustrated in Chapter 5 of this book. The doctrine of nihilism is the great theology and the spiritual treasure of the religion of Cao Daiism and Buddhism.

Halos and radiant light particles formed an immense tricolor celestial ring, which surrounded the extreme light shaped like a marvelous lotus and located exactly at the center of the primitive universe. The tricolor celestial ring had three colors: exceptionally clear and extremely bright yellow on top, the transparent greenish blue in the middle, and the crystal clear purplish red at the bottom; and nobody could understand the mystery of that tricolor celestial ring formation and its color representation.

After several years of research, Julie and I found that the tricolor represents the universal color of creation and also it is the physiognomy of the birth of God Cao Dai/Jehovah. We also found that the Cao Daiism Flag does have three colors as previously stated; the transparent bright yellow on top, the clear greenish blue in the middle, and the crystal clear purplish red at the bottom. Both Buddhism and Cao Daiism did assure that the giant extreme light (*Thái Cực Đăng*) was the primitive Almighty God's Supreme Spirit as God Cao Đài/Jehovah said in the book of the selection of *The WORDS of GOD* through great mystery as Reference No. 1 shown on page 1, and the tricolor celestial ring was exactly Almighty God's Holy physiognomy (*Thiên thể*) and it was known that God Cao Đài/Jehovah visited Temple Lama in Tibet once in every fifty years to teach religion. Whenever God Cao Đài/Jehovah visited this temple, His Holy physiognomy lit up the entire Temple Lama and made it look marvelous and, of course, prettier than sunlight.

Some Tibetans did see the beautiful Holy physiognomy of God Cao Đai/Jehovah that no words could be used to describe correctly the beauty of His Holy physiognomy.

A lot of people have the same question—how fast can God Cao Đai/Jehovah travel in this Universe? It was known that a few seconds after praying for God Cao Đai/Jehovah to teach religion at the Tòa Thánh Tây Ninh in Viet Nam or at the Temple Lama of Tibet, God Cao Đài/Jehovah was there immediately.

God Cao Đài/Jehovah uses the space compact link mystery to travel around the Universe at the velocity of thought. The distance from the Kingdom of God Cao Đài/Jehovah located near the Polaris to our Planet Earth No. 68 is approximately seven billion light-years;

Almighty God takes seconds to travel that distance and light has to take nearly seven billion years to travel. One space compact link mystery is approximately equal to one trillion light-years with no limit regarding higher super eminently supernatural God's supreme power. In the spiritual world, the supernatural extreme intelligence requires to possess the super ultra tangible physiognomy or holy body to metamorphose internal energy into supreme power or to transform natural energy (similar to dark energy or nuclear energy) into real power for all supernatural activities, similar to human beings' need for oxygen, water, and food to transform them into living power. The supernatural intelligence alone cannot create a giant universe without metaphysical power.

God Cao Đài Tiên Ông Đại Bồ Tát Ma Ha Tát/Jehovah did have two bodies, the tricolor celestial ring is God's Holy body or physiognomy and the extreme light is God's supreme spirit or this was the first logo.

Time Began

At the present, the Universe and God Cao Đài Tiên Ông Đại Bồ Tát Ma Ha Tát/Jehovah are approximately fourteen billion years old.

Before moving to the North Pole, God Cao Đài Tiên Ông Đại Bồ Tát Ma Ha Tát/Jehovah integrated a large amount of nuclei left over by the first thermonuclear fusion explosion, *Nổ Ầm* or Big Boom, and metamorphosed them into metaphysical power. Almighty God solely utilized His internal power to force nucleons in nuclei to pair together for gaining extremely supernatural power; this mystery needs more scholarly research to bring about the truth. It is pretty certain that nuclei in some way related to God Cao Dai/Jehovah's supreme power. With this power, God Cao Đài Tiên Ông Đại Bồ Tát Ma Ha Tát/Jehovah could create, size, build, and position any constellation, galaxy, super nova, black hole, and star, etc., in the Universe.

After integrating enormous power, Almighty God moved to the North Pole of the Universe. Almighty God who had peerless and unparalleled power, and who was certain self-existing and ever existing, spent most of His time and energy thinking how to shape up the Universe and created heirs.

The Beginning of Time

Once again:

Reference 1:
The selection of the Words of Almighty God (Book 1), page 31; Thursday, July 22, 1926. The 'Great Mystery' was from Almighty God Cao Đài/Jehovah teaching religion for the Third Amnesty in the Orient. "The Nihility Gas gave birth to one and only Thay (your master). Who gave birth to others such as Confucius, Buddha (Siddhartha), Lao-Tse, Jesus Christ, etc…? It was from Tao (way or religion). All my children need to understand. If Thầy (your master) does not exist, thereby nothing will take place in this Universe; if there was no Nihility Gas thereby Thầy (your master) did not exist either." The Words of God Cao Đài/Jehovah proves the Truth.

Reference 2:
"Electron Density and Temperature Measurements in a Low Pressure Magnetoplasma" by Hue Le at the Tennessee Technological University, August 1976.

Note: Cao Đài Tiên Ông Đại Bồ Tát Ma Ha Tát is the Holy name that God Jehovah proclaimed during teaching religion in the Orient for the Third Amnesty.

The Purpose of This Book

The purpose of *The Beginning of Time* is to prove the truth of how God Jehovah/ Cao Đài Tiên Ông Đại Bồ Tát Ma Ha Tát was born by His own words. (The nihility gas gave birth to one and only Thầy, your master, if Thầy did not exist thereby not thing will take place in this universe, if there was no Nihility Gas thereby Thầy, your master, did not exist either) and the birth of the Universe by the Plasma Theory in conjunction with first thermonuclear fusion explosion, *Nổ Ầm* or Big Bang (Big Boom).

Since the truth is here, so the road from the truth to the beatitude will be naturally connected with obvious certainty, and people will no longer take for granted the beatitude to search for the truth (God) anymore.

In addition, for more than a hundred thousand years, God Cao Đài/Jehovah did send a lot of His messengers and great philosophers to teach His doctrines on Earth. The great theologies that brought peace and love to our society are:

- Buddha Siddhartha with the philosophy of philanthropy
- Confucius with the philosophy of charity and happy medium
- Lao-tse with the great philosophy of Taoism (honest and impartial)
- Jesus Christ with fairness and justice

The most popular of the social conventions are benevolence, righteousness, politeness, virtue, and trust. If everyone on Earth shares a small practice of charity, philanthropy, Taoism, fairness, and justice, then the whole world would be in peace overnight and our society would turn into a paradise shortly thereafter; love would come to everyone.

This is also the truth (the birth of the Universe and the birth of God Cao Đài/Jehovah by His own words) for which all human beings had been searching for a long time.

Since the Stone Age (approximately 100 thousand years) people accepted beatitude to search for the truth (God); now the truth is found and obviously certain so people will take the road from the truth to the beatitude and not to be in reverse.

There Is No End of Time

Since the whole Universe was originally created from only hydrogen gas and nothing else, the first thermonuclear fusion explosion, *Nổ Ầm*/Big Bang, was processed by a proton-proton chain reaction, which was the process of two protons smashing each other and fusing; this process gave birth to deuterium, neutrino, and anti-matter positrons. As stated above, the first thermonuclear fusion explosion, *Nổ Ầm*/Big Bang was the gigantic fireball that's overall diameter was approximately equal to four billion light-years across and full of protons (hydrogen).

Immediately after deuterium was created, a nearby proton came to fuse with it and formed helium-3 and a gamma particle was also created. Now helium-3 was plentiful in the gigantic fireball and they came smashing into each other and the product of this process was helium-4 and two protons (hydrogen) were reborn again.

This process alone tells us that the primarily primitive source of life is hydrogen, which could be utilized to create all of Nature and the whole Universe; it also recreates its own existence.

In addition, hydrogen atoms are plentiful outside and inside of our universe to support Nature forever; in other words, hydrogen is the primary source of creation and hit also recreates its own existence. The primitive sources (hydrogen) of creation, per se, will ever last; therefore, time never ends.

Conclusion: The primary source of life never ends, so time never ends, either.

The nihility gas was the formation of the matter electron and anti-matter positron; they smashed and annihilated each other (voided), and radiant light particles were emitted so-called extreme light (*Thái Cực Đăng*) or God Cao Đài/Jehovah's Supreme Spirit. Matter electrons and anti-matter positrons will be plentiful and last forever.

Eternal life is also defined from here. All things in the Universe will never be destroyed; they only change phases along with the course of Nature.

Please read the doctrine of nihilism in Chapter 5 for more information.

Outside of Our Universe

God Cao Đài/Jehovah utilizes the number twelve (12) very often to build the Universe and Nature; the little universes, which are for the most part vertebrates, including human beings; and the thoracic cage has twelve pairs of ribs supported by twelve vertebrae. The Universe has twelve signs or constellations, which support life for all Nature; all religions have had twelve permanent theological masters to teach their Gospels or philosophy, and God Cao Đài/Jehovah has twelve supernatural timer divines to control the twelve signs or constellations to always be at the exact positions in the zodiac circle to protect the life of all Nature. (Please see Chapter 3 for more details.) The number twelve is the key setting number to be utilized to build inside and outside the Universe.

Our universe is one among twelve other universes, which are revolving on one plane circle called the super zodiac. The radius of the super zodiac is equal to twelve space compact links. Again, one space compact link is approximately equal to one trillion light-years. The supernatural God built the super universes, which is a huge sphere with the overall diameter approximately equal to twenty-four space compact links.

The super zodiac and the super universe have the same center. The super zodiac is divided into twelve equal parts; each part takes up thirty degrees even. The distance from center to center of all universes is approximately equal to six and one-fifth space compact links. Some of these universes are probably still empty and God Cao Đài/Jehovah wants His children to continue His good works of building in the empty ones like the Universe where we live now.

We take some more time to go farther beyond the super universe; we will see the greater super universe, which has twelve super universe and the radius of the greater super universe is double in size, which is equal to twenty-four space compact links. The farther we go, we will see the greater, and greater, and greater super universe; the work of the supernatural God will never end.

Chapter 2

Mystery of the Creation

At first, God Cao Đài Tiên Ông Đại Bồ Tát Ma Ha Tát/Jehovah was the peerless beautiful and intelligent lotus of the primitive universe who had unparalleled and matchless supernatural power, the first logo, so-called Brahman. Brahman was the highest rank in the Spiritual World or Buddha. After disintegrating His own supernaturalness, God Cao Đài/Jehovah coalesced it with a giant amount of the universal cold matter, which was the majority of electrons left over by the *Nổ Ầm* or Big Boom and the dark energy for the second logo of Creation called Civa, who was the supreme negative creator of the Universe; all followers of the Cao Dai religion worship Civa as the mother god or goddess. God Cao Đài Tiên Ông Đại Bồ Tát Ma Ha Tát/Jehovah set up the first positive and negative principle.

All natural living things created or that will be created must obey the positive and negative principle to exist; otherwise, it would be destroyed or become nonexistent. Civa, the second logo of the Creator of the Universe or mother god (goddess), had utilized the positive and negative law and mysteriously combined with the first logo's power to create the third logo, which included one great male creator called Great Fohat, one little male creator called Little Fohat, one great female creator called Great Koilon, and one little female creator called Little Koilon or Trinity. The Trinity later obeyed the laws and principles of God Cao Đài Tiên Ông Đại Bồ Tát Ma Ha Tát/Jehovah and built the Universe (Cosmos). The mother god has total power and or authority to create tangible life and nature. God Cao Đài Tiên

Ông Đại Bồ Tát Ma Ha Tát/Jehovah had the supreme power to evaluate and change everything with respect to His will and principles or presetting models.

The Sequence of the Creation:

1. The virgin nihility gas gave birth to extreme light (*Thái Cực Đăng*) or God Cao Đài/Jehovah, the first logo.

2. God Cao Đài Tiên Ông Đại Bồ Tát Ma Ha Tát/Jehovah disintegrated His own supernaturalness and coalesced with the primitive universe cold matter and matter electrons for the second logo or Civa (mother god), who belonged to the negative side of the universal creation.

3. The positive and negative principle was created by God Cao Đài Tiên Ông Đại Bồ Tát Ma Ha Tát/Jehovah. Mother god utilized this law by combining her power source and that of the first logo to give birth to the third logo or Trinity (one great male and one little male called Fohat and one great female and one little female called Koilon).

4. Supernatural Creator or the third logo obeyed the positive and negative principle and the will of God to flourish the eight diagrams of prosperity. Finally the Universe started to flourish from here.

The Kingdom of God Cao Đài/Jehovah

In order to maintain the Universe functioning and prospering correctly, God Cao Đài Tiên Ông Đại Bồ Tát Ma Ha Tát/Jehovah gave orders to the third logo to create twelve Supernatural Time Commander Divines to carry out His will and principles to protect life in the Universe. These twelve Supernatural Time Commander Divines had metaphysical power strong enough to carry out all commands assigned by Almighty God, including the building and positioning of any constellation, galaxy, supernova, black hole, and star etc., in the Universe. Each Supernatural Time Commander Divine has full authority to monitor and control each sign of the zodiac; in other words, they can reposition any constellation at any time to maintain the exact time preset by Almighty God's scheme to protect life and nature. The mother god (second logo or Civa) and twelve Supernatural Time Commander Divines collected hydrogen, deuterium, helium, plasma products of different kinds, and nuclei left over by the first thermonuclear fusion explosion, *Nổ Âm* or Big Boom, to build the North Star (Polaris) approximately one degree away from the North Pole.

As the next step, mother god and the twelve Supernatural Time Commander Divines built another Polaris companion star, which was rotating and revolving around the Polaris. This Polaris companion star was much smaller than the Polaris so that we cannot see it with our naked eyes. The Kingdom of Almighty God was built on the giant white crystal float that normally travels around the Polaris companion star in such a way that those two stars take turns shining in the Kingdom of Almighty God at all times without interruption of light; the white crystal float that carries the Kingdom of Almighty God Cao Đài/Jehovah could exit its orbit and travel anywhere around the Universe at any time.

If anyone utilizes the vocabularies to describe Paradise in describing the Kingdom of Almighty God, he/she will be totally wrong. In Paradise crystal waters are falling from a mountaintop a thousand feet high and winding around mountain brooks; running waters are sparkling and playing miraculous music as if all visitors are under illusion. Multicolor flowers are open to welcome the spring morning and boast their beauty under a blue sky like a million golden angels in a lost sea. Singing birds are sending us to dream of rest that never ceases. Far away mountains are rising up endlessly and make us feel like we're in a lost world. Gourmet fruits are hanging on green trees, running as far as you can see. Winding trails are running along blue lakes as far as the horizon; schools of gold fishes are upstream to enjoy the fresh water this spring can offer. This wonderful world sends you to dream in an endless place of joy and peace, and makes you forget to return.

The Kingdom of Almighty God was built with white crystals, sapphires, diamonds, and many precious stones for conducting and filtering high energy light from Polaris. There are two places where God Cao Đài Tiên Ông Đại Bồ Tát Ma Ha Tát/Jehovah often presides:

1. The forbidden golden gated kingdom palace is an endless place of giant monuments, splendid palaces, and magnificent towers as far and as tall as your eyes can see; these charming wonders do not exist anywhere in the Universe. All these wonders were built with diamond-spars, crystal clear sapphires, and different kinds of rare precious stones, making us fall into this virtual image and giving us mystically feelings. Right at the center of this wonderful place, a tall golden gate encrusted with golden dragons opened up to the forbidden kingdom palace of God Cao Đài Tiên Ông Đại Bồ Tát Ma Ha Tát/Jehovah.

2. Passing through countless wonders and magnificent monuments, splendid highways embedded with white crystals and curbed with sparkling sapphires take us to a mystical city that mushrooms up with solemn temples as far as the distant horizon. Nobody could tell the reality of these highways at all because traveling on these highways would give everyone a feeling like walking on halos or on the marvelous mysterious air of fairyland. The mysterious structure of these temples make you feel as if the high ranking spirits are haunting there at all times, but they are not there. Passing through these mysterious temples is an immense sparkling lake scattered with inconsistent lotus of white, red, and yellow; a white crystal road takes us to the tallest and largest temple where God Cao Đài Tiên Ông Đại Bồ Tát Ma Ha Tát/Jehovah is teaching religion and commanding the Universe.

The Nihilism Statement in God Cao Đài Bible (*Ngọc Hoàng Kinh*)
The Explanation of Nihilism

A similar structure to this temple can be found on Earth, which is Tòa Thánh Tây Ninh located in Tây-Ninh Province in South Việt-Nam. This Tây-Ninh Cao-Đài Temple was built with Earth materials by the prophet of the Cao Đài religion, Mr. Phạm Công Tắc, through the great mystery and/or meditation. There was no architect or engineer to do any work on this temple. After a long time of daily meditation, the Prophet received a small drawing of a portion of the whole project sketched by Đức Lý Thái Bạch, who was the very high-ranking Buddha in the Kingdom of Almighty God Cao Đài/Jehovah. A large number of the Cao Đài followers came and built that portion that day; the next day the Prophet Phạm Công Tắc had to meditate again to obtain the next sketch for the next day's work. The construction of this Cao Đài Temple in Tây Ninh, South Viet Nam, took at least several years to complete as we see it today. This wonderful Temple of Cao Đài Tây Ninh is an exact copy of the Throne of the God Cao Đài/Jehovah in Mahaparanirvana and this temple represents the God Cao Đài/Jehovah's kingdom on Earth. It is a must see in order not to be surprised when anyone comes to see God Cao Đài/Jehovah after Earth life.

One important thing all spiritual dignitaries need to remember is that all super highways, connecting routes, and bridges in Mahaparanirvana or in Nirvana are built on air and connected together by extremely bright halos in such a way that all dignitaries who practiced charity, philanthropy, fairness, and justice during life on Earth would travel on them without fear; otherwise, fear would push them off and they would fall into the Dark Sea of Sin for reincarnation, obeying the law of Karma. Their astral bodies would be soaked with dark water of shame and become ugly. Most of the honest religion builders and their honest followers who carried out the will of God Cao Đài/Jehovah by practicing charity, philanthropy, fairness, and justice are welcome to this marvelous fairyland in Nirvana and comfortably enjoy their past humanitarian practices. The top secrets of success of these honest religion builders could be found in Confucianism, Christianity, Taoism, and Buddhism. Looking in all directions at this place, everything seemed to conceal supernatural power and mystery always dominated our respectfully serious feeling.

TOA THANH TAY NINH

The Beginning of Time

After the completion of the Kingdom of Cao Đài Tiên Ông Đại Bồ Tát Ma Ha Tát/Jehovah, the mother god (*Phật Mẫu*) and the twelve Supernatural Time Commander Divines (*Thời Thần*) spent their time and energy on building the zodiac, the Southern Cross, constellations, galaxies, stars, supernovas, and black holes, etc. in the Universe with respect the pre-designed models and structures of God Cao Đài/Jehovah.

Mother god (*Phật Mẫu*) totally holds the authority of building tangible lives and all other material structures in the Universe. God Cao Đài Tiên Ông Đại Bồ Tát Ma Ha Tát/Jehovah only checks, evaluates, and changes everything with respect to His will and principles.

The long road that every soul must go through from material soul, vegetation soul, animal soul, human soul, deity soul, saint soul, angel soul, and Buddha soul, which is the highest rank in the spiritual world.

Reference:
Thánh Ngôn Hiệp Tuyển, quyển nhứt, trang 68.

Translation:
The great mystery written by Almighty God (*Ngọc Hoàng Thượng Đế Viết Cao Đài Giáo Đạo Nam Phương*) on Sunday, December 19, 1926.

Thầy (your master) is the name that Almighty God proclaims when teaching religion or speaking through great mystery.

All my children listen. There is one thing that all of you never know about; comprehension of religious doctrine is extremely important and admirable. It will explain why everyone must cultivate and purify oneself. All of you were born on this planet, Earth. You live, go through a lot of hardship, and finally die in this world.

Thầy (your master) asks all of you: What will happen to you after your death?

Where are all of you going? None of you could understand the supernatural answer. Thầy (your master) explains: The metempsychosis's mystery changes material soul into vegetation soul, vegetation soul into animal soul, animal soul into human soul, and human soul into highly spiritual dignity, etc; and, at last, into Buddha.

The secret of metempsychosis that allows material soul to evolve into vegetation soul, into animal soul, into human soul, into highly spiritual dignitary, etc., and, at last, into Buddha…is the Evolution/Creation.

Human beings have had to devolve thousands, even millions of times to become pious dignitary. Pious dignitaries are classified into many elevations. You are on Planet Earth No. 68, the King on this Planet Earth is not even qualified to be the lowest dignitary on the Planet Earth No. 67. On Planet Earth No. 67, pious dignitaries are also classified into many elevations. The elevation of dignitaries will be promoted higher and higher until reaching Planet Earth No. 1. The next steps will be through the four giant quadrants of the Universe: One quadrant is reserved for evils or transgressors to cultivate and purify themselves to regain their previous spiritual elevation, the other three quadrants are administered by three prophets separately.

Then all of you must go through three thousand worlds. The three thousands worlds are also built with the solar systems like our Planet Earth and the Sun, the dignities on three thousands worlds are also required to be cultivated and purified to earn higher spiritual elevation.

After the three thousands worlds, the route will take you to the thirty-six heavenly skies. In the thirty-six heavenly skies, all spiritual dignitaries have only two bodies—the second body ('holy' body), and the third body (the Soul); their worlds are built in the air without a solar system. All spiritual dignitaries in the thirty-six heavenly skies must continue to cultivate and purify to reach the First Heavenly Sky.

At the first heavenly sky, all spiritual dignitaries are well cultivated and extremely purified to be qualified to reach the Kingdom of God Cao Đài Tiên Ông Đại Bồ Tát Ma Ha Tát/Jehovah or *Niết-Bàn* (Nirvana). In the Kingdom of Almighty God located near Polaris and its companion star, the so-called *Niết-Bàn*, which was built on a giant white crystal float mentioned above, all spiritual dignitaries are ranked Buddha; God Cao Đài Tiên Ông Đại Bồ Tát Ma Ha Tát/Jehovah is the highest rank of Buddha.

There are several billions people on the surface of Planet Earth No. 68; all of them have a unique form and were created by a similar principle, but different colors listed from black, gray, green, red, yellow, and white.

The pious dignitaries were created by evolution-creation, which means that the evolution of all living creatures on this planet Earth allows change with respect to the will and principles of God Cao Đài Tiên Ông Đại Bồ Tát Ma Ha Tát/Jehovah; for example, people born in the twenty-first century A.D. are better looking and more intelligent than the people who were born hundreds of thousands of years B.C.

At the present time, planet Earth is going through the third manvantara of the Almighty God creation. Each manvantara has thirty-six thousand years and is divided into three cycles. Each cycle has twelve thousands years.

The first (upper) cycle of any manvantara has a higher percentage of pious dignitaries and/or prodigies; the percentage of pious dignitaries of the second (middle) cycle is far less than the one of the Upper Cycle; and the evils are outnumbered by the righteous people in the third (bottom) cycle. People of the bottom cycle must go through a lot of challenges and testing of hardship to be qualified in reaching higher elevations of pious dignitaries for the next planet Earth (Planet Earth No. 67), otherwise (failure of testing) they must remain on this planet Earth and become the race of lucidity. By the prediction of the Prophet of the Cao Đài religion, Đức Hộ Pháp Phạm Công Tắc said that Planet Earth No. 68 will be cleaned out to make room for the next generation of supernatural and prodigy people. All the people who worship Satan and demons will be destroyed and irretrievable. Currently, all living creatures on our planet Earth (No. 68) are at the end of the third (bottom) cycle of the third manvantara and will enter the first or upper cycle of the fourth manvantara. There will be a large number of highly spiritual dignitaries among the children of God Cao Đài Tiên Ông Đại Bồ Tát Ma Ha Tát/Jehovah; more children of God Cao Đài/Jehovah will share and practice His will: charity, philanthropy, fairness, and justice.

Chapter 3

Number Twelve Is God's Own Special Number

Reference 1: (See explanation on page 1.)

Reference 3:
Thánh Ngôn Hiệp Tuyển, quyển 1, trang 12.
"Thập Nhị Khai Thiên là Thầy, Chúa cả Càn Khôn Thế-Giới; nắm trọn Thập Nhị Thời Thần vào tay. Số mười hai (12) là số riêng của Thầy."

Reference 3 Translation:
The collection of the words of God Cao Đài Tiên Ông Đại Bồ Tát Ma Ha Tát/Jehovah, Book 1, page 12.
"I, Thầy (your master), am exactly the founder of the twelve signs of the sky, the Master of the whole universe; the Commander of the Twelve Supernatural Timer Divines. The Number Twelve (12) is my Own Special Number."

The History of the number 12 is as follows:
Before building the Universe, God Cao Đài/Jehovah set up the standard for measurement. God Cao Đài/Jehovah went back to the atomic structure, which is the primitive source of life and looked at the atomic diameter, which varies greatly due to ambient temperature; in other words, when an atom absorbs energy, all its electron clouds are moving outward from its small central region known as the core or nucleus. God Cao Đài/Jehovah went deeper inside the

— 12 —

Thập nhị khai Thiên là Thầy, Chúa cả Càn Khôn Thế Giới; nắm trọn Thập-Nhị Tuổi Thần vào tay. Số mười hai là số riêng của Thầy.

... Chưa phải hồi các con biết đặng tại sao vẽ Thánh Tượng «Con Mắt» mà thờ Thầy, song thầy nói sơ lược chổ biểu chất-ảnh.

Nhãn thị chủ Tâm,
Lưỡng-quang Chủ-tể,
Quang thị Thần,
Thần thị Thiên
Thiên giả, ngã giã.

Thần là khiếm-khuyết của cơ mầu nhiệm từ ngày Đạo bị bế. Lập «Tam Kỳ Phổ Độ» nầy duy Thầy cho Thần hiệp «Tinh-Khí» đặng hiệp đủ «Tam Bửu» là cơ mầu nhiệm siêu-phàm nhập Thánh.

Các con nhớ nói vì cớ nào thờ con mắt Thầy cho chư đạo-hữu nghe...

... Phẩm vị Thần, Thánh, Tiên, Phật từ ngày bị bế Đạo, thì luật-lệ hỡi còn nguyên, luyện pháp chẳng đủ, sang Thiên-Đình mỗi phen đánh tắt «Thần» không cho hiệp cùng «Tinh Khí».

Thầy đến đặng huờn-nguyên Chơn-Thần cho các con đắc đạo. Con hiểu «Thần cư tại Nhãn». Bố trí cho chư đạo-hữu các con hiểu rõ. Nguồn cội Tiên Phật do yếu-nhiệm là tại đó. Thầy khuyên con mỗi phen nói Đạo hằng nhớ đến danh Thầy.

13 Mars 1926.
**NGỌC-HOÀNG THƯỢNG ĐẾ VIẾT CAO-ĐÀI
GIÁO-ĐẠO NAM-PHƯƠNG**

Thầy cho các con biết trước, đặng sau đừng trách rằng quyền-hành Thầy không đủ mà kềm thúc trọn cả Môn Đệ.

— 81 —

Như kẻ bên Phật-Giáo hay tặng Nhiên-Đăng là Chưởng-giáo; Nhiên-Đăng vốn sanh ra đời Hiên-Viên Huỳnh-Đế.

Người gọi Quan-Âm là Nữ Phật-Tông, mà Quan-Âm vốn là Từ-Hàng Đạo-Nhơn biến thân. Từ-Hàng lại sanh ra lúc Phong-Thần đời nhà Thương.

Người gọi Thích-Ca Mâu-Ni là Phật Tổ, Thích-Ca vốn sanh ra đời nhà Châu.

Người gọi Lão-Tử là Tiên Tổ-Giáo, thì Lão-Tử cũng sanh ra đời nhà Châu.

Người gọi Jésus là Thánh Đạo Chưởng-giáo, thì Jésus lại sanh nhằm đời nhà Hớn.

Thầy hỏi vậy chớ ai sanh ra các Đấng ấy ?

Khí Hư-Vô sanh có một Thầy. Còn mấy đứng Thầy kể đó ai sanh ? Ấy là Đạo. Các con nên biết.

Nếu không Thầy, thì không có chi trong Càn-Khôn Thế-Giới này; mà nếu không có Hư-Vô Chi-Khí, thì không có Thầy.

Dimanche 25 Juillet 1926. (16-6-B.D.),

CAO ĐÀI

Cười . . . .

T . . . ., con coi mặc Thiên-Phục có xấu gì đâu con ?

Một ngày kia sắc-phục ấy, đời sẽ coi quí-trọng lắm. Con ôi ! con có biết những điều ấy bao giờ !

Cười . . . .

Mấy đứa Lễ-Sanh có, dọ sắp-đặt sự nghiêm trang trong đàn cầu Thầy, chớ chẳng phải dạy đủ đi lễ mà thôi, mỗi đại đàn phải đủ mặt; chúng nó phải ăn-mặc trang-hoàng hai đứa trước, hai đứa sau, xem sắp đặt sự thanh-tịnh. Thầy dạn các con như dạn nội chẳng nghiêm. Thầy không giỡn, ba con nhớ nghe !

Tr . . ., J . . . K . ., T . . ., nghe;

atom and found that the nucleus of any atom remains constant in size and very stable. God Cao Đài/Jehovah took the diameter of a nucleus and set it as a unit length as follows:

$$\text{Unit Length Lu} = \text{Nuclear Diameter Dn}$$

God Cao Đài/Jehovah set the number 12 for the practical length of measurement called M.

$$M = Lu \times 100 \text{ trillion}$$
Or:
$$M = Dn \times 100 \text{ trillion}$$

Obviously, M is now the standard length so-called 'meter' being used worldwide. If we divide the standard length M (meter) by 100, we will get 1 centimeter or cm.

$$CM = M \text{ divided by } 100 = Lu \times 1 \text{ trillion}$$
Or:
$$Lu = Dn = 10^{-12} \text{ cm}$$

The nuclear diameter was chosen by God Cao Dài/Jehovah. $Dn = 10^{-12}$ cm as the measurement standard unit length.

Everything on our planet Earth or located anywhere in the Universe was created or will be created by His genius. Children must obey His laws and principles.

It is obviously God Cao Đài/Jehovah who chose the nuclear diameter for the primitive measurement standard His own special number 12 for that purpose.

God Cao Đài/Jehovah created Supernatural Timer Divines for controlling and managing all the changes and movements of everything in the Universe that could slip away from His preset schemes or standards. Before creating life in the Universe, God Cao Dai/Jehovah built the zodiac to support life based on His own special number 12.

The zodiac did not naturally occur, but rather built by the preset model of God Cao Đài/Jehovah. It is a band of the celestial sphere, extending about eight degrees to either side of the ecliptic; it is also the pathway of the sun, the moon, and the principle planets. This band is divided into 12 equal parts called signs; each part is exactly 30 degrees wide and carries the name of a constellation or sign; this mysteriously secret construction of the zodiac with respect to the Number 12 proves to us that God Cao Đài/Jehovah is really self-existent and ever-existence. Each sign takes up 30 degrees; 12 signs will take 360 degrees to complete the whole circle of the zodiac. An elliptical or circular diagram represents this band containing pictures of animals, human figures, etc. that are associated with the constellations. All Nature needs to understand, the principle planets, the moon, and the sun appear to move in a large circle, passing through the same sign (constellation) over the course of one year. The sun passes through one constellation per month as Earth revolves around it. This system was adapted worldwide and later most countries in every corner of the world independently recognized a similar system. A year is divided into 12 months for four seasons. It is obvious that four seasons support the production of food, experience, and good prediction of repeating movements of the sun, principal planets, stars, and the Moon will be the key of survival. In other words, collecting experiences of these phenomena will greatly benefit all of Nature's existence and prosperity.

Supernatural phenomena that command our life and change our world are from Almighty God Cao Đài/Jehovah. Lightning creates fire as our gift; wind distributes vegetation seeds around the world for food, soil conservation, and many living necessity materials; and rain contributes greatly in farming and supports lives.

Going back to the ancient time, there was a Jewish folklore named Golem when God Cao Đài/Jehovah tried to artificially create a human being endowed with life by supernatural means. The foot of the Golem was divided into twelve equal parts called inches. The foot and inch system does exist and is still applicable around the world at present time.

That Golem could not advance and prosper; God Cao Đài/Jehovah changed His mind and created cells for His evolution-creation of all living creatures in the Universe.

In Christendom, Jesus Christ had twelve apostles to preach his Gospel. In Mormon society, the Church has twelve members of the Administrative Council. Cao Đài religion has Thập Nhị Thường

The Beginning of Time

Quân, so-called Twelve Permanent Theological Masters. Confucianism has Twelve Top Philosophy Masters. Taoism has Twelve Top Philosophy Masters. Buddhism has Twelve Top Disciples to teach the Buddhism Doctrine. Normally at the farmer's market, people exchange ears of corn by the dozen or 12, which is from the primitive life standard. God Cao Đài Tiên Ông Đại Bồ Tát Ma Ha Tát/Jehovah was the founder of the zodiac or the first twelve signs (constellations) of the sky as previously stated.

There were twelve spiritual elevation zones in the Universe:

Three Blue Azures:
    1. The 'Giant Blue Azure' also called Mahaparanirvana.
    2. 'Upper Blue Azure' also called Anupadaka or Paranirvana.
    3. 'Crystal Blue Azure' also called Atma or Nirvana.

After the completion of the three blue azures, the third logo took the order of God Cao Đài/Jehovah to build three more Upper Boundary Worlds:

    1. The World of Angels or Spiritual World of Buddha is divided into three elevations:
        A. Upper
        B. Middle
        C. Lower
    2. The World of Superior Mentality is divided into three elevations:
        A. Upper
        B. Middle
        C. Lower
    3. The World of Mentality (the Tangible World) is divided into three elevations:
        A. Upper
        B. Middle
        C. Lower

There are a total of nine Upper Boundary Worlds.

The lower elevations are the worlds, which have solar systems, including our seventy-two Planet Earths and the other 3,000 worlds in the Universe with solar systems. In these worlds human beings, animals, and vegetations live together like our Planet Earth No. 68.

Three Blue Azures and nine Upper Boundary Worlds make twelve Spiritual Elevation Zones in the Universe.

God Cao Đài Tiên Ông Đại Bồ Tát Ma Ha Tát/Jehovah built the vital portion of human being's body, which is the thoracic cage with twelve vertebrae and contains twelve pairs of ribs supported by twelve vertebrae.

The similarity between the Universe and Nature is the Universe has twelve signs or constellations, which revolve harmoniously with the suns, the living planets, and the moons, etc.—the so-called thoracic cage of the Universe to support sensitive life and all of Nature. The center of intelligence is the Kingdom of God Cao Đài/Jehovah, which is located near the North Star; there are trillion of trillions of deity super highways connected between the Kingdom of God Cao Đài/Jehovah and the Southern Crux.

The thoracic cages of all vertebrates including human beings also have twelve pairs of ribs to protect sensitive organs and support life. The central intelligence is the brain, located on top of the head and the key communication between the central intelligence and the lower part of the body is the spinal cord. God Cao Đài/Jehovah built the Universe and all vertebrates, which are little universes in a unique structure and similar formation of the number 12 as cited by His own words. Our four fingers have total of twelve finger-joints per hand, excluding the thumb.

Chapter 4

The Constitution of Trinity in Unity (The Three Bodies in Human Beings)

Since God Cao Đài Tiên Ông Đại Bồ Tát Ma Ha Tát /Jehovah used trial and error on how to create his children to continue his good work of building a new universe.

First, God Cao Đài Tiên Ông Đại Bồ Tát Ma Ha Tát/Jehovah artificially created a human being endowed with life by supernatural means called Golem, which was the name of Jewish folklore being well known until today. That artificial human being can live normally, but he cannot advance or prosper at all. God Cao Đài/Jehovah tried different ways to create His children by creating the cells of different kinds to form different organs and, at last, the constitution of a human being was achieved. Cells were classified in numerous kinds from vegetation, insect, and animal to human being. Cell is the smallest structural unit of an organism that is capable of independent functioning, consisting of one or more nuclei, cytoplasm, various organelles, and inanimate matter, all surrounded by a semi-permeable plasma membrane. The structure of the cell is very complex and concealed in a supernatural mystery, never be discovered.

God Cao Đài/Jehovah dismantled His own supernatural abilities, the invisible necessity of his intelligence source, and honored it to every living cell with different function and responsibility preset by Almighty God to perform thinking and all physical activities necessary for prosperity and protecting life. Since cells are the smallest organism, they need clean food and require not to be abused by drugs, alcohol, polluted environment, etc. to live up to their life expectancies. Cells are medical doctors, warriors, engineers, builders, producers, repairers, information carriers, and the loyalist slaves. For example, an animal was injured by a hunter or fighting with other animal to survive; all the cells associated with the wound would continuously work to repair the damage and provide essential chemicals to prevent infection and stop bleeding, as well.

When the wound is completely healed, all the cells would stop the healing task and go back to their guard or caretaker duties as usual. They do not overwork or perform insufficient care, and they are ready to serve at all times. These inborn responsibilities of all living cells are from God Cao Đài, preset in the intelligence spot of every cell so that scientists cannot copy or duplicate it.

From vegetation cells to human being cells, Almighty God carefully assigned limitations of all individual functions and responsibility; scientists have never discovered the mysterious secret of Almighty God Cao Đài/Jehovah. The brain cells of human beings are far more complex and concealed a lot of supernatural power utilized in spiritual communication between the great Universe and human beings, they can transform high frequency dimension messages from deity world into low frequency dimension information or perform psychiatric functions to care for psychosis. The spiritual worlds, including the thirty-six heavenly skies and the Kingdom of God Cao Đài Tiên Ông Đại Bồ Tát Ma Ha Tát/Jehovah are the worlds of super high frequency dimension; all 3,072 worlds with solar systems, which include our Planet Earth No. 68 are the worlds of low frequency dimension. Human beings who reside on the 3,072 planet earths have three bodies in one: The first body is the flesh (tangible) body or corporeal; the second body is the semi-tangible body or corposant, which is also called the holy body or astral body; the third body is the supernatural intelligence spot, which is also named the soul.

The higher elevation of spiritual dignitaries who reside on the thirty-six heavenly skies and on the Kingdom of Almighty God do not have the first body (corporeal), they only carry the second body (holy body) after millions of times of changing and purifying through their metempsychosic lives and the third body, or the supernatural intelligence spot, which is the soul.

The holy worlds of the thirty-six heavenly skies are built in the air without solar systems and are positioned close to the Kingdom of God Cao Đài Tiên Ông Đại Bồ Tát Ma Ha Tát/Jehovah.

The brain is a giant supernatural computer and sized smaller than a pumpkin; as said above it is extremely complex and concealed a lot of mysterious secrets that scientists cannot copy or duplicate. To simplify the communication between the brain and the Universe, God Cao Đài Tiên Ông Đại Bồ Tát Ma Ha Tát/Jehovah created the zodiac to maintain life and prosperity of all Nature. Changes in the Universe are indirectly signaled to all of Nature and our lives are normally affected; prediction is our survival skill.

The high frequency dimension messages are sent from high elevation of spiritual world to human beings through the cerebral cortex and received by the thalamus; this route is called the 'Deity Superhighway.' The thalamus then sends this message to the cerebellum where the high frequency dimension message will be transformed into low frequency dimension information.

The cerebellum will create a respondent message and return it through the spinal cord to the heavenly worlds; this route is called the 'Heaven Canal.' The highly purified or devout people could achieve this communication pathway or the zodiac in humans after many years of practice. Similarly in the zodiac, the Deity Superhighway is from the constellation of Cancer down to Sagittarius, and the Heaven Canal is from Capricorn up to Gemini. The doctrine of Lao-tse also illustrated that the thalamus and the cerebellum together form a zodiac in human beings, and the fontanel of human beings is the essential rendezvous between Almighty God Cao Đài Tiên Ông Đại Bồ Tát Ma Ha Tát/Jehovah and human beings. The fontanel is a sacred place where the prophet presides and keeps evils away from us.

After successfully creating cells for everything from vegetation, insects, animals, and human beings, etc., God Cao Đài/Jehovah created all living beings with a unique form which contains head, body, and limbs (in different forms), but having the same utilization for daily activities to prosper.

The first body is a tangible body and requires food and different living necessities to flourish; the tangible body is also called corporeal or flesh body. When the first body eats food, the purity of food will be transformed into blood. A portion of the blood will carry nutrients to nourish all the cells of the first body; the rest of the blood will evaporate into gas and escape the first body through the fontanel. This gas will cover the first body and mold to it exactly, the same figure called the corposant or the second body; some people also name it as the holy body or the astral body. The second body (holy body) is semi-visible, very lightweight, and superbly electrically conductive. Since the second body is from the first body, similar to the water taken from the well or from the river, if the well or the river is dirty or polluted, thereby the water will be dirty or polluted, as well. Likewise, if the first body is impure or sinful, then the second body or astral body will be similarly impure or sinful. The impurity of the first body means that the corporeal is contaminated by killing animals and eating their flesh, alcohol, prostitution, drug addiction, inhumane abuse, or by any activities that are against Nature—evil activities, etc. It is important to understand that animals are one step behind human beings; they love their offspring better than human beings do, and never kill their own children for honor in any situation. They always take good care of their young. In danger, animals normally fight to the death to protect their offspring. Some people are worse than animals; they kill or sell their own children to satisfy their ill will or for material benefit.

Inhumane activities and good works are always recorded by the corposant from the beginning to the end of everybody's life. Only Satan or demon worshippers are the people who kill their own children in cold blood for stupid reasons. They try to steal the children of Almighty God and turn them into evil for killing purposes. Killing is the great sin and should never be forgiven.

The second body (corposant) has an important relationship between the first body and the third body, which is the soul. It is the extreme intelligence supernaturalness spot disintegrated from God Cao Đài/Jehovah and is given to every human being; it is unbreakable and indestructible by any means. When God Cao Đài/Jehovah gives His extreme light of intelligence spot to anybody or all of His children; He never takes it back and it is for them forever.

The third body (soul) is a tiny bit or spot of intelligence that God Cao Đài/Jehovah disbursed from His supreme spirit and contributed to all His children. God Cao Đài/Jehovah endows immense power to every second body/astral body to protect life for all living nature.

This Holy power may evolve by cultivating, purifying, and fulfilling Almighty God's will to achieve an infinite spiritual level that is invisible to scientists.

Fulfilling Almighty God's will is to practice charity, philanthropy, fairness, and justice. These three great philosophies could be found in Confucianism, Buddhism, Christianity, and Taoism; they are also the highest mysterious secret of success for the honest religion builders.

Knowing the relationship between the **three bodies** in human beings is very important. The second body (corposant/holy body) is the essential medium of the three bodies; it always listens and carries out all the requests of both the first body and the third body, or the soul.

In the tangible world, the first body (corporeal) normally requests to be well fed and demands more luxury material for comfortable living; by that reason, the second body/holy body has to work hard to meet the demand and gradually distances itself from the wisdom of the third body.

The second body (corposant) from time to time ignores the spiritual requests from the third body/soul and dedicates itself to searching for more material to please the first body by any means—even killing, pillaging, or performing inhumane activities to spread ill will.

Since the relationship between the first body and the second body is similar to the water taken from the well or from the river; if the well or the river is polluted or impure, therefore, the water will be polluted or impure. Likewise, if the first body/corporeal is impure or sinful, then the corposant/astral body will be impure or sinful because the astral body is directly from the first body, similar to the water taken from the well as stated above. The relationship between the second body/corposant and the third body/soul is far more important than that of the first body/corporeal.

God Cao Đài Tiên Ông Đại Bồ Tát Ma Ha Tát/Jehovah disseminates His own supernaturalness ability and bestows it on all His children and all living nature. The enormous powerful spot of intelligence could be cultivated and purified to achieve the higher and higher spiritual elevation close to Buddha's power and invisible to science.

The combination of the holy body and the spiritual power could transform all ionized gases (plasma form) into extreme energy to deal with the tangible world; beside integrating nuclear power, all supernatural divines and goddesses normally exercise that power to build constellations, galaxies, black holes, super novas, and stars, etc. throughout the Universe. Satan, the son of Almighty God, did have that power; however the intelligence spot that God Cao Đài Tiên Ông Đại Bồ Tát Ma Ha Tát/Jehovah disbursed His own supernaturalness and endowed it to him and to all his children like a grain of sand compared to the desert as a whole.

In order to get rid of Satan, and satanic and demonic worshippers, Almighty God just took a quick look at them, they were destroyed instantly. Almighty God does not want to do so because He wants to exercise charity, philanthropy, fairness, and justice to all nature and His children. These evils continue to run deeper into opposing Almighty God and steal Almighty God's children for demonic purposes; their sins will be grave and receive deadly punishment afterward. Their races will disappear; their astral bodies will be destroyed, and their souls will lose spiritual power causing it to be irretrievable.

Once again, God Cao Đài Tiên Ông Đại Bồ Tát Ma Ha Tát/Jehovah created all kinds of cells and bestowed them with the necessity of intelligence for advancement and prosperity.

After the unsuccessful creation of Golem, Adam was the first white human being that God Cao Đài Tiên Ông Đại Bồ Tát Ma Ha Tát/Jehovah made with advanced cells and granted him knowledge, humanitarianism, philanthropy, charity, fairness, and justice. Adam was formed in the image of God Cao Đài/Jehovah. God Cao Đài/Jehovah made male and female; the male was formed first and then the female, with respect to His positive and negative principle. Adam was placed with Eve in the Garden of Eden to till it and keep it in order; however He violated God's rules. On his transgression, a sentence of death was passed upon him, and finally he was expelled from the Garden of Eden. Afterward he had children—Cain, Abel, and Seth when he was 130 years old. Adam was at last dying at the age of 930. The creation of cells and forming human being were the first successes of God Cao Đài/Jehovah after many trials and errors.

Since the first body/corporeal is the Almighty God's temporarily instrument to produce and/or to transform material into semi-tangible gas to form the second body/corposant, which is necessary to carry the most precious gift from God Cao Đài Tiên Ông Đại Bồ Tát Ma

Ha Tát/Jehovah, the supernatural intelligence spot, or everybody's soul.

There are four essential constituents in all living beings and nature, as follows:

1. Tangible Constituent (*Vật-Thể*)
2. Ether Constituent (*Khí-Thể*)
3. Sacred Constituent (*Thần-Thể*)
4. Eucharist Constituent (*Thánh-Thể*)

Summary of the Four Constituents:

1. The first tangible constituent is the human being (corporeal), which is the combination of earth, water, fire, and wind. The corporeal is enveloped by an extremely light ether constituent called the vital spirit (*Phach* or *Linga-Sharira*).

2. The ether constituent is a near-zero mass medium, so-called vital spirit, which envelopes the corporeal. With the support of the magnetic field, the vital spirit provides energy and power for the corporeal to exist and to protect the corporeal from falling apart and decaying; in other words, it protects life.

    Whenever the magnetic field accidentally is disconnected from the corporeal by magic power, drugs, or by the end of life, the ether constituent will be separated from the corporeal and the corporeal will be motionless and unfeeling like a dead body. The naturally clairvoyant or psychic people can see the ether constituent when it is detached from the corporeal; it looks like a dim mist shaped as a human. Whenever the magnetic field is disconnected from the corporeal and cannot be reconnected, all parts of the body falls apart and decays or death occurs.

3. The sacred constituent is also called Kama or astral body, which is formed by the evaporation of blood escaping the body through the fontanel and molding the corporeal or

copying the corporeal. The sacred constituent is connected to the ether constituent by a magnetic field.

The sacred constituent plays the role as a medium between the corporeal and the soul; it is also a viscera, which stores desires and sentiment. In other words, it is the source of feeling whenever it is stirred up; the eyes love to see beauty, the nose loves perfume, the ears love sounds of music, tongue loves taste, and the body loves affection, etc. It is the primary motivation of the six roots of sensation (eyes, ears, nose, tongue, body, and mind). The Kama of highly purified and cultivated people could leave their corporeal and travel through 3,000 worlds in a second (much faster than the speed of light) to contact with divines and spirits. The Kama could convert dark energy into active energy and utilize it to transport the soul in motion or transform it into extremely magical power for spiritual works, which are beyond scientist's knowledge. The sacred constituent also carries the psychological name of 'consciousnesses' and it is the essential medium of the tangible constituent and the soul.

4. The Eucharist constituent is also called inferior mental corps, which carries the psychological name 'intellectual'; it is a communicator between the Kama (sacred constituent) and the soul. Whenever the sacred constituent justifies right from wrong, good from bad, or moral from evil, then the Eucharist constituent will follow that judgment and notify the human soul for comprehensive reasons. In short, every human being has four constituents: tangible constituent, ether constituent, sacred constituent, and Eucharist constituent. Tangible constituent and ether constituent are associated with our tangible world; sacred constituent is homogeneous with the deity world, and the Eucharist constituent is in the same family with sainthood.

Every human being has three Souls: human soul *(nhon hon)*, spiritual soul *(anh hồn)*, and divinity soul *(linh hồn)*. The human soul is equivalent with the mentality of Earth life; spiritual soul is also called intellectual or sainthood soul; and divinity soul is equivalent with Buddha, which is Crystal Blue Azure or Nirvana.

The upper blue azure and the giant blue azure are the forbidden zones for highest spiritual elevation where God Cao Đài/Jehovah and supreme spirits normally reside. Once again, there are twelve spiritual zones: three blue azures, three spiritual worlds or Buddha, three worlds of superior mentality (the middle world), and three worlds of mentality or tangible worlds. The twelve spiritual zones are exactly preset models of the Universe of God Cao Đài/Jehovah stated above.

The Destination of Human Souls after Death

Death is simply considered as when the magnetic field that binds the tangible constituent of a corporeal and its ether constituent together is disconnected. The ether constituent is the vital storage of energy to support life; if it is separated from the corporeal, the corporeal cannot keep going on without energy and gradually falls apart or dies.

Within approximately a week or so after death, the ether constituent is completely separated from the sacred constituent and becomes less active; it hangs around the dead corporeal all day and at night it moves to a quiet area, such as an abandoned home or a graveyard and waits for the corporeal to completely decay. At the time that the corporeal is completely decayed, the ether constituent dissipates in the air immediately. People who have the sense of clairvoyance can see the ether constituent and call it a ghost. Cremation of the dead body may be a way to help its ether constituent to dissipate faster.

When the human soul breaks away from the corporeal its ether constituent feels like a prisoner waiting to be freed from jail; the human soul now unloads the burden, which is the stinky and dirty corporeal and becomes more active. The human soul now will live normally in the astral world and does not feel tangible life anymore because of its disconnection from the material world. At this time, the human soul is still under the protection of the sacred constituent or Kama; the Sacred Constituent always envelops the human soul, similar to the ether constituent that envelops the corporeal.

The sacred constituent is homogeneous with the astral world; in other words, it is at home and to be considered as a means of transportation to carry the human soul in this astral world. The sacred constituent (corposant) continues to utilize dark energy to transport the human soul in the astral world with history recorded during its lifetime. It is important to know that the corposant and the astral world are of the same matter.

At the astral world (so-called Hell or the Inferno), the lifetime records will be replayed from the beginning to the end of life like a movie; all the activities from good to bad, from noble to evil, and from spiritual to non-spiritual are displayed clearly and in good order. The human soul will comfortably enjoy all activities exercised—charity, philanthropy, fairness, and justice; otherwise it will sinfully suffer because of ill will, immoral practices, cruelty, and inhumane deeds. Comfortable joy (beatitude) and sinful suffering are the spiritual reward and punishment, respectively, in the astral world for all activities done during the tangible lifetime. The human soul of righteous people and people who sacrificed their lives to protect their country will receive spiritual awards in the astral world and be honorably promoted to deity. In the Astral World, all the human souls have the same level of intelligence and always help each other advance to the higher spiritual elevation. The long or short duration in the Astral World will depend on the good and the bad record of a life. When the human soul is liberated from the sacred constituent, it will dissipate in the astral world and all the lifetime incarnation records will be transferred to the Eucharist constituent for future metempsychosis.

The holy body of the human soul now has only one Eucharist constituent and becomes lighter and freer to move to the top spiritual elevation of the Astral World, which is the World of Inferior Mentality or Sainthood World. In the Sainthood World the incarnation records will be justified again; if the records are good, then the human soul will have the sainthood promotion and enjoy beatitude for a long period of time, otherwise life metempsychosis will take place by the law of Karma.

When the limitation of time staying in the Sainthood World comes to an end, the Eucharist Constituent will dissipate in this world and the human soul will ascend to the World of Superior Mentality, or Paradise, which is the Angel World. The duration of the stay in Paradise will depend on how good or excellent were the practices of charity, philanthropy, fairness, and justice while on earth.

The human souls who practiced God's will are allowed to stay in Paradise for a long time to enjoy beatitude; when the duration stay reaches the expiration date, the human souls have to return to tangible life, taking their place anywhere among 3,072 solar systems like our solar system for carrying out God Cao Đài/Jehovah's will for higher spiritual elevation.

On the way back to tangible life, human souls start to receive Eucharist constituent in the World of Inferior Mentality and the sacred constituent in the astral world. When the human souls are reaching the tangible worlds, which is one of the 3,072 worlds with solar systems, the human soul will receive the ether constituent and the physical body from new parents. The human soul will stay in this world to carry out the expected works as assigned while in Paradise.

Normally the devout human souls will unload the heavy burden including the physical body, ether constituent, sacred constituent, and Eucharist constituent at once and ascend to Nirvana or the World of Buddha.

Chapter 5

The Doctrine of Nihilism

The Beginning of Time

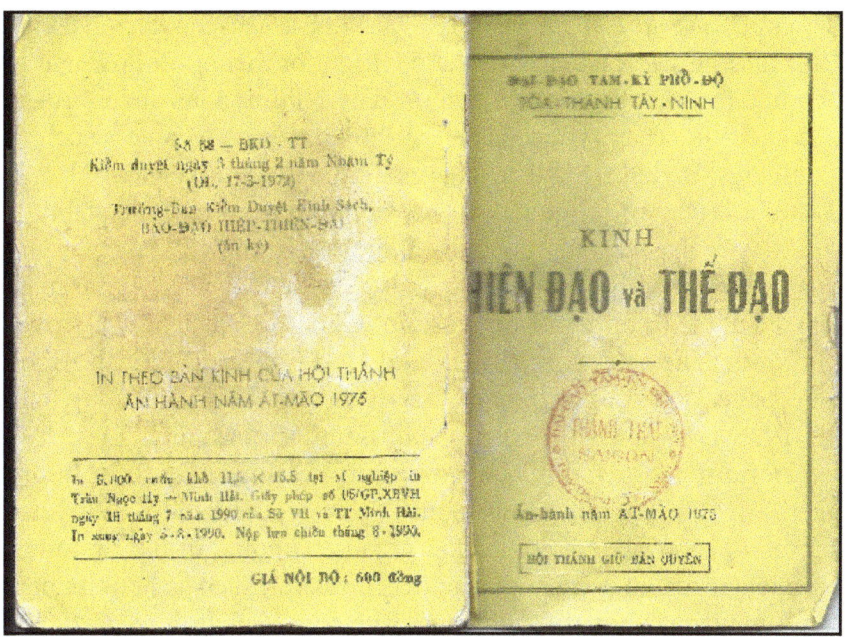

Ngọc Hoàng Kinh (God Cao Đài Bible)

God Cao Đài Tiên Ông Đại Bồ Tát Ma Ha Tát/Jehovah did create His children; He also created the doctrine for them to follow and to practice His theology. Once again, "The Nihility Gas gave birth to one and only Thay (your master). Who gave birth to others such as Confucius, Buddha (Siddhartha), Lao-Tse, Jesus Christ, etc…? It was from Tao (way or religion). All my children need to understand. If Thầy (your master) does not exist, thereby nothing will take place in this Universe; if there was no Nihility Gas thereby Thầy (your master) did not exist either." The Words of God Cao Đài/Jehovah proves the Truth. The selection of Words of God Cao Đài/Jehovah, Book One, page 31, through *Great Mystery* (Huyền Cơ) on Thursday, July 22, 1926 at Ngọc Đàn, Cần Giuộc of South Viet Nam.

Definition of Nihility Gas: Approximately one hundred millions years after the *Nổ Ầm* or Big Boom/Big Bang, the primitive universe started to clear up and cool off, a gigantic amount of the first thermonuclear fusion explosion products left over, such as helium, protons, neutrons, positive and negative ions, subatomic particles, matter or first electrons, and anti-matter or first positrons, etc., could be found abundantly in the primitive universe. The matter electrons and anti-matter positron (the formation of nihility gas) were smashing each other (Be Existent or Yes Means One = 1), annihilated each other, or rendered each other void (Be Nothing or No means Zero = 0) to emit radiant light particles or photons ($e^-e^+ \rightarrow \gamma^* +$ energy). This process kept growing enormously gigantic and rapidly self-built an extreme light shaped like a pretty lotus, which was God Cao Đài/Jehovah's Supreme Spirit; in other words, the nihility gas gave birth to the one and only God Cao Đài/Jehovah exactly as God Cao Đài/Jehovah said when He taught religion for the Third Amnesty through *Great Mystery* (Huyền Cơ) on Thursday July 22, 1926 at Ngọc Đàn Cần Giuộc, South Việt Nam.

In the Bible of God Cao Đài/Jehovah we can find these words on line number seven and on line number nine from the top of the "Book of KINH THIÊN ĐẠO và THẾ ĐẠO," Page 22, line one thru line three from the top of this page: line one reads, "*Nhược Thiệt, Nhược Hư*"; line two reads, "*Bất ngôn nhi mạc Tuyên dại hóa.*" and line 3 reads, "*Thị Không, Thị Sắc.*"

Translated, line 1 reads, "Be existent, be voided," and line 3 reads, "Be voided, be existent," similar to the binary digits 0 or 1.

Line 2 reads, "Quietly the law of Nihilism (Yes = 1 and No = 0, or vice versa) established the immensely giant universe with respect to the preset model of God Cao Đài/Jehovah."

The "Be existent, be voided," and "Be voided, be existent" phenomena were repeated over and over again. The repetitive activities of the nihility gas that gave birth to God Cao Đài/Jehovah will never cease, which means that God Cao Đài is self-existence and ever-existence.

The existence of matter electron and anti-matter positron that they encountered annihilated each other or rendered each other void, then void will be replaced by their next existences again and again… matter electrons and anti-matter positrons could be found abundantly in the Universe at all times.

There are two primary applications of the Doctrine of Nihilism:

1. Physical Science -

The 'Be Voided, Be Existent' conceals the meaning of 'Voided or Naught = Zero (0), Existent or Yes = One (1)' or similar to the binary digits 0 and 1. These binary digits 0 and 1 are heavily applied worldwide in our daily life.

As we observe everything around us daily from our wristwatch, microwave oven, radio, television, coffeemaker, car dashboard, game machine, digital camera, cell phone, children's toys, laptop computer, super computer, highway traffic controller red lights, public security control systems, air traffic controller, International Space Station, Hubble telescope, nuclear power station control system, hydro electrical generation control systems, super jets, hospital radiology systems, the complexity of the television network, the audio and video telephone systems, etc., these systems could not be built without the binary digits 0 or 1; they are extremely necessary for our daily life even required to save life, to maintain security for society, to keep the world in order regarding communication, transportation, entertainment, natural disasters, world war, terrorists, etc.…

The human brain normally has approximately 110 billion neurons. Neurons are the basic units of the central nervous system; they are processing data, transmitting electrical signals (electrical impulses), and communicating with each other through synapses. Neural networks consist of many thousands of neurons and they are connected with each other via synapses. Normally the dendrites of the neurons receive the nerve signals and the axons of the neurons send them onward to the required destinations. The active parts of the neurons can produce multi-millions of molecules, which are recognized as the smallest units containing a couple or more atoms, so-called neurotransmitter molecules. The active part of a neuron, called soma fabrications neurotransmitter molecules, which are tiny bits of different chemicals; after fabricating these tiny bits are enveloped in thin film marble-like accumulators. These accumulators, called synaptic vesicles, are traveling in a unique direction toward the pre-synaptic membrane like a freight train; when the synaptic vesicles reach there, they will softly fuse with the pre-synaptic membrane and give up all the neurotransmitter molecules. Finally the neurotransmitter molecules are free to go through the gap between the pre-synaptic membrane and the postsynaptic membrane, the so-called synaptic cleft.

When these neurotransmitter molecules reach the postsynaptic membrane, they enter their receptor site, which is the tiny cavity located on top of each sliding door of the postsynaptic membrane.

The postsynaptic membrane has many double sliding-door gates, each having a small cavity equal to the size of the neurotransmitter molecule on its top. When the neurotransmitter molecules are in their cavity, they release their chemical to force the sliding doors to open and build an electrical pathway through the postsynaptic membrane. This narrow pathway only allows the positive charge particles from the synaptic cleft to pass through and nothing else. The electrical information signals will be formed by zero current (Zero = 0) and the spike of the amplitude of the positive charge particle (Yes = 1) only. The new binary coding waves of information (including audio and both color or black-and-white video signals) generated by different sliding doors are encoded immediately and recorded throughout the brain for future recall and decoding. This process will continue until the input information is discontinued. This biological structure of the human brain can be found in neurology books or in the brain cell or neuropathology books, etc. In the near future, scientists could discover a super human brain-like artificial intelligence to take over some parts of a damaged brain to save lives. God Cao Đài/Jehovah had been using the Doctrine of Nihilism a long time before human beings discovered computer chips and it is still on high demand today.

2. Supernatural Science -

The Doctrine of Nihilism has another meaning in the supernatural world and affected our supernatural life, as well. God Cao Đài/Jehovah provided material, which is necessary for us to live and enjoy life during the time on Earth; it will come and go along with the world changing and the time passing by.

Many kingdoms on earth encompassed people, nations, luxury materials, and some space in the Universe (Be Existing = Yes or 1) for centuries; the whole world changed with the passing of time, and some kingdoms disappeared, as well (Rendering Void or Nothing Left = Zero or 0).

In the world, many people were born rich and many people were born poor; the rich people owned a lot of luxury things such as castles, mansions, luxury yachts, gold, diamonds, limousines etc., and the poor people owned some torn blankets, tattered clothes, poor meals, some used hats, and steep highway overpasses as their best shelter. All of them were happy with what they had, but after a generation or so, none of them could be found; the rich people and the poor people were changing in the same way—returning to dirt—and the volume of space their bodies took up on Earth before (Be Existing = Yes or 1) were now be empty (Be Voided = Zero or 0).

One time the rich folks owned luxury things and the poor people owned near trash materials (Existing Material or Yes = 1); none of these things would last forever. All the torn blankets and luxury materials were gone together (Be Voided = Zero or 0) as their bodies were decaying. These empty spaces once occupied by older generations (Be Voided = Zero or 0) would be replaced by the next generation (Be Existing [again and again] = Yes or 1); the new fortune will be built again and again, after all, everything will disappear again and again. These recurrences of lives and materials signify that everything that occurred in our lifetime could be temporary and must obey the law of Nature to be cycled and recycled.

Every one of us should understand that during the lifetime on Earth, someone loved materials and killed others to satisfy their ill will, or for plundering wealth (Existing Material = Yes or 1) etc.,

One day he or she returns to dirt, his or her hands will be empty (Empty = 0), but the sin of killing and pillaging is still recorded in his or her astral body for future judgment executed by the Court of Heaven. All the luxury materials and poor belongings are temporary that God Cao Đài/Jehovah provided for us to enjoy to live and to do good things for mankind and shortly they would be gone. Nothing on Earth will belong to anyone forever; they come and go as if they were under the control of Nature and no one can change their course of disappearance (Nothing Left = Zero or 0) and existence (Reappear = Yes or 1); the Doctrine of Nihilism conceals the power of Nature that God Cao Đài/Jehovah preset for all living Nature. A billion years ago the earth was shaped differently than today; continents and mountains that once stood strong, have now disappeared or turned into oceans, and oceans turned into continents.

In other words, the children of God Cao Đài/Jehovah do not use materials to build their 'Eternal Life' they are just temporary and disappear shortly after life; the real building blocks to be utilized to build eternal life are charity, philanthropy, fairness, and justice. If the population of the whole world shares in the practice of charity, philanthropy, fairness, and justice; peace and love will come overnight for everyone. Once again, God Cao Đài/Jehovah created his children and expected them to take after His good works of building a new universe. He wants his children to be purified and sin free, and be deserved to succeed to His throne or to be the new master of the new universe.

A human being has three bodies—the first body is the corporeal from his or her parent, and the Second Body is the corposant, which is from the first body; in other words, when we eat food, the purity of food turns into blood, part of the blood will carry nutrition for all parts of the corporeal, some blood will evaporate into the gas form and leave the body through the fontanel.

After getting out of the first body (corporeal), this gas will copy the first body exactly like molding it for future reorganization. It is composed by three things—*Tinh*, *Khí*, and *Thần*.

The Beginning of Time

The Origin of Three Bodies in Humans:

1. From God Cao Đài/Jehovah:
   Ngọc Đế sinh quang vận bát hồn
   Càn Khôn hóa trưởng tại Chí Tôn
   Huệ Văn Lê

2. From Mother God or Second Logo:
   Vạn vật thân vi từ sinh khí
   Diêu Trì Kim Mẫu tạo hữu vi
   Julie Phan Lê

Translation:

1. From God Cao Đài/Jehovah or the First Logo:
   God Cao Đài creates eight souls from His own supernatural power (the giant extreme light) gratifies and transposes metempsychosis to all living Nature. The Universe is growing maturely under the control of the most Venerated Supreme God Cao Đài/Jehovah.

2. From Mother God or the Second Logo:
   All tangible lives and Nature are originated from living gas (hydrogen). Mother God creates all tangible existences.

Tinh is our tangible body from our parents, which were from Mother God's creation and also called the first body.

Khí is a purified gas or evaporated blood and originated from the mystery of Mother God also called the second body.

Thần is the extreme light or God Cao Đài/Jehovah's Supreme Spirit, which is disintegrated from Him and bestowed to His children and all Nature also called the third body or the soul.

Tinh, *Khí*, and *Thần* are required for all living creatures; three become one and one becomes three. During the lifetime on Earth the second body (*Khí*) was always around us to protect our lives and it is also an essential communicant between the first body and the third body (the soul). The third body (the soul) alone will not be qualified to present in front of God Cao Đài/Jehovah after life; the relationship between the second body and the third body is extremely important, similar to the space shuttle and the astronaut. The space shuttle cannot fly without the astronaut; likewise, the astronaut alone cannot fly either. Whenever life expectancy ends, the second body will leave the dying body and directly coalesce with the third body (the soul). If the second body is purified and sin free, the human soul will ascend to a higher spiritual world; otherwise, the second body has to stick with the decaying body (the first body) and death will capture it. The Cao Đài religion has a Bible to retrieve sinful human soul and liberate the heavily sinful astral body from the deceased body for reincarnation. The third body (the soul, Thần) requires it to join with the second body (the astral body, Khí) to return to the astral world, otherwise impossible.

The second body needs to be lighter than the air so it can lift and carry the soul, which is the third body, to the next holy world or to the new spiritual elevation even to present in front of God Cao Đài/Jehovah every human soul required having a sin-free astral body. The best way of purifying oneself is not to kill animals and eat them; practice charity, philanthropy, fairness, and justice or be a devout person to fulfill the will of God Cao Đài/Jehovah, and the roadway to beatitude will be at hand.

Since the second body is semi-tangible, lightweight, semi-visible, and an excellent electrical conductor, if it is unpurified, very sinful, and loaded with killing, cheating hypocrisy, pillaging, inhumane practices, deceitfulness, and immorality, etc.…all the wealth gained from unrighteous means will be gone after life; but its sins were accumulated, recorded, and weighing heavily on the astral body so it cannot leave the stinking body or get out into the air and will have a good chance of being struck by lightning because the astral body is a super electrical conductor, and it is an irretrievable destruction.

The second body does have a copycat brain of the first body. All the information that was recorded by the first body's brain will be transferred to the copycat brain of the second body for future judgment executed by the Court of Heaven in the astral world; the history of every human soul will be replayed like a movie from the beginning to the end by the reflection of halo. During life on Earth, if anyone practiced charity, philanthropy, fairness, and justice, he or she will enjoy receiving a delightful award and ascend to Heaven; otherwise, he or she will suffer severely of shame. It is self-punishment and the life recurrence will be affected by the law of Karma.

It is, above all, the top mysterious secret of success of the honest religious leaders is humanitarian practices and exercise of charity, philanthropy, fairness, and justice. These moral philosophies could be found in Confucianism, Buddhism, Christianity, and Taoism.

Chapter 6

The Will of God Cao Dai/Jehovah

The Will of God Cao Đài Tiên Ông Đại Bồ Tát Ma Ha Tát/Jehovah teaches:

1. Charity
2. Philanthropy
3. Fairness and Justice
4. Practicing Social Conventions

The world that we live in has different kinds of people of different spiritual levels; the first one probably was the metempsychosis from rats, snakes, pigs or other animals; the second one may be the return of the previous human beings by the law of Karma; the third one may be high-ranking spiritual divines who voluntarily descend to earth to save children of God Cao Đài/Jehovah; and others may come back to earth to pay back what they owed from a previous life, some of them probably are the people who loved their material things so they keep hanging around their home and/or grave in the afterlife.

The knowledge or comprehension levels of these people are totally different. Overall, after the first manvantara (the first 36,000 years), a large number of human souls successfully became high spiritual dignitaries. During the second manvantara (the second 36,000 years), less than half of human souls successfully became high spiritual dignitaries. During the third manvantara (the third 36,000 years), the evil were outnumbered by the righteous human souls.

Each manvantara has three cycles: the upper cycle, the middle cycle, and the bottom cycle; each cycle has 12,000 years. At the present time, we are in the bottom cycle of the third manvantara and God Cao Đài/Jehovah wants to shape up our world for entering the upper cycle of the fourth manvantara. God Cao Đài/Jehovah prepares the holy route for His children of self-improvement and purifying to achieve spiritual life during the Third Amnesty.

On my left palm God Cao Đài/Jehovah had shown clearly three Pyramids side by side and mounted on the head line similar to the three pyramids built at Giza in Egypt more than 5,000 years ago. The largest pyramid at the far left represents our Galaxy, the middle pyramid is medium size and has a Star on top of it representing our sun, and the last pyramid on the right is the smallest one, which represents our Planet Earth No. 68; on top of the pyramid representing our planet Earth stands a secular tree, which has three big branches. The first branch represents charity found in the philosophy of Confucius, the second branch represents philanthropy found in Buddhist philosophy, and the third branch represents fairness and justice found in Christianity and Taoism. My left and right palms each show a mystic cross; this means that God Cao Đài/Jehovah bestows me with the gift of clairvoyance to discover the universe and the spiritual world.

I asked God Cao Dai/Jehovah whether or not 'Doom's Day' would happen during the galactic alignment on December 21, 2012. The answer was, "Doom's Day will never happen, but the evils will gradually be destroyed."

All satanic and demonic worshippers and some races will actually disappear forever. God Cao Đài/Jehovah wants to clean out our planet Earth and build a new society for the supernatural and prodigy people.

Since the Stone Age, the children of God Cao Đài/Jehovah had advanced and prospered rapidly so, God Cao Đài/Jehovah wanted to bring new and up-dated religion for His children with old and new philosophies and new ways of life. The most effective way of life is to re-unite key philosophies of all religions that were always the principles that God Cao Đài/Jehovah wanted all His children to practice and live in harmony with God Cao Đài/Jehovah among them, but away from evil.

God Cao Đai/Jehovah did not ask His children to do anything difficult, simply:

1. Obey Nature
2. Respect Natural Prosperity Principles (Positive and Negative Law)
3. Practice Moral and Courteous Social Conventions
4. Practice Charity, Philanthropy, Fairness, and Justice

If everyone in the whole world shared in the practice of these principles, peace would come overnight and there would be love for everyone shortly thereafter.

The Dangerous Addiction of Alcohol and Narcotics

The general belief is that all diseases are caused by viruses, bacteria of all kinds, abuse by all means, or insufficient sanitary arrangements, etc.... However, abuse of alcohol and narcotics is also a major cause of disease.

As specified in Chapter 4, cells are the smallest organism, and they need clean food and require not to be abused by drugs, alcohol, or polluted environment, etc....to live up to their life expectancies. Cells are medical doctors, warriors, engineers, builders, repairers, information carriers, and the loyalist slaves. Normally the heart pumps the dirty blood that returns from the body to the lungs for cleaning and purifying; the lungs require some time to clean, purify, and enrich the blood before pumping it back to nourish all the cells in the body.

Alcoholic people may get drunk very often; similarly drug-addicted people may frequently get high also. Whenever anyone gets drunk or high on narcotics, his or her heart pumps at an abnormally fast speed, so the dirty blood being is pumped to the lung for cleaning and purifying, but the lung does not have time to do so.

The dirty blood is immediately pumped back to the body without purifying and enrichment. Year after year of this cleansing failure and all the cells of his or her body become sick and weaker and weaker due to the polluted blood; they lose their defense power, maintenance capability, and other holy functions given by God Cao Đài/Jehovah. Finally, his or her body will lose all immunities; this is the main cause of cancer and the invasion of other diseases.

The fontanel is the holy rendezvous of God Cao Đài/Jehovah and human soul; it is also the place where the prophet presides to keep all evil away from us. When a person gets drunk, the fontanel will widely open making room for evil to invade and the prophet will go away; the person will lose consciousness and control, as well. Since evil takes over the brain, it will dominate that person to commit evil acts such as killing, and become involved in immoral or lawless activities, etc. Drinking a little alcohol for digestion now and then is forgivable, but drinking to get drunk is dangerous and unacceptable. The best thing is to never start drinking or doing drugs; live a healthy life and stay away from deadly suffering caused by alcohol and dangerous drugs.

Chapter 7

Intelligent Versus Genius and Allocation of Wealth

Intelligent People

Since the birth of mankind, people try to measure the immeasurable knowledge and set an imaginary scale to allocate wealth with respect to people's labor, skill, and comprehension. Some people could process addition, multiplication, or divisions faster than a calculator; however, they could not repair a wristwatch, fly a space shuttle, or sing a song to make a living. Others know how to sing songs and become very wealthy; many people know how to repair cars, build houses, pave the roads, or take care of the sick and old people, but they only make a bare living. People keep searching for a scale that could bring equality and balance to society, but there is nothing in sight. The inequitable and injustice always dominate our material world, the powerful people cheat the powerless people, and the word equality is just a sound of nonsense or just an artificial sugarcoating.

After several years as a freedom fighter, I was off and on, in and out of the jungle of Viet Nam fighting against the communist Việt Cộng, where life and death was around the corner at all times and the jungle diseases alone could kill anyone at anytime. I was fortunate to get out of that deadly war alive under the protection of God Cao Đài/Jehovah and had a chance to go back to college to gain higher education.

South Viet Nam was lost to communist North Viet Nam amid my pursuit of education; I had to work two hours a day with the payment of one dollar and twenty-five cents per hour as a janitor to support my education until I received a fellowship for a Ph.D., majoring in Engineering Science, with payment of six hundred dollars per month. During that time my wife, Julie Phan Lẻ, and my son Phong Bằng Lê were left behind in Viet Nam under the cruel Communist Regime. Neither my wife nor my son ever had a minute of peace during that time and wanted me to get them out of Viet Nam; the communists punished them with hardship and chased them wherever they were hiding.

I had to sacrifice my education and look for a job to get my family out of the cruel communist regime as soon as possible. I did have a job at a Southern company; I was always mentally suffering because of my family left behind in Viet Nam and the scar of the deadly war embedded deeply in my life, making my blood pressure extremely high and causing nervous tension all the time. I kept conquering myself to overcome these difficulties because my family needed help from me and only me.

I worked very hard through the international government in Europe and the government of the USA and, after nearly nine years of separation, my family arrived in Belgium on March 29, 1979 and reunited with me on August 21, 1979 in Alabama, USA.

For nearly one year, I was dedicated to learning and working very hard; I helped the group to finish the Unit 1 Miller Steam Plant of the Alabama Power Company. Time was changing quickly; nuclear engineering was in high demand, and Electrical Director John Plaxco and Nuclear Manager William Garner requested that I transfer to the Nuclear Support Department. I did have strong background in electrical engineering, instruments and controls, and nuclear safety codes and standard design; after nearly four years of hard work, I helped to complete the design of a large nuclear power plant, the Alvin W. Vogtle Unit 1 PWR (Pressurized Water Reactor). I did a lot of good work on nuclear design and support; my job did not stop there, again, I was transferred to the new nuclear support department at E. I. Hatch. My knowledge and skill with nuclear safety codes and standards, electrical system control, and instrument and control design were growing rapidly, and I created the new philosophy for nuclear support and technical design.

First, all designs must be completed with state-of-the-art Code of Federal Regulations (10CFR, parts 0-199), and standards must be applied without short cut, and be clear and informative.

Second, all designers, drafters, and engineers must consider that the design must contain complete clear information, and it is the one and only instrument to be utilized as a communicator between the architect engineer and site engineer to save trips to the plant.

Third, the most important and obligatory information required for shutdown must be available at all times on the design, regardless of repeating information; this could help operators to take action quickly to save the plant in a dangerous mode of operation without searching all over for instructions. Operation power and control power are highly recommended in system coordination to avoid unnecessary shutdown caused by minor faults to save the plant for long life and reliability. All vital equipment control power was required to be equipped with a UPS backup system to avoid hot shutdown caused by minor loss of control of power supply; for example, loss of control power of main turbine-generator, etc.... The UPS backup power systems are not too expensive; however, they could provide a high standard of operation reliability and save plant life from hot shutdowns caused by minor troubles.

Fourth, all equipment mounting details (small or large) are required to meet seismic standard application or at least using the civil safety code.

Fifth, technology is changing every second, so all people who work on nuclear design or support are required to be up to date and learn new skills.

Sixth, to be sincere and honest; people can lie and cheat their family and friends, but not in nuclear services.

Seventh, management of a nuclear power company requires having knowledge of nuclear technology; promoting an idiot to handle CEO, EVP, VP, manager, and supervisor duties, etc., would be very dangerous for nuclear decision making and disrespected by others. Making the wrong decision or being a 'yes-man' only creates difficult problems and chaos among employees; this could bring failure to the company.

Promotion of CEO's, managers, or supervisors who do not have any background with nuclear technology will create deep troubles, even making it impossible to do the right thing in the nuclear services. I left the company for that reason; they were just tumbling blocks of the work force and making it difficult, even impossible, to carry out the right nuclear services.

The Beginning of Time

The Ugly Supremacy of Power Versus Money and Racism

During twenty-six years of nuclear service, I have experienced a lot of ugly things that happened to my life.

I tried to attend all the meetings called by the big shot CEO of the Southern Nuclear Operating Company, Mr. H. All I learned were just the big words, "Look at the mirror," A thousand times he said this and never told us for what. None of the meetings mentioned anything about technical services or advanced nuclear technology at all.

The words "look at the mirror" reminded me that I knew one guy who looked at the mirror more than ten times a day to comb his hair and check his sharp clothes to have affairs with many married women; this guy was later was killed by a 122-millimeter rocket that left no remains to be found. Another guy also looked at the mirror more than ten times a day to see how he could build pretty muscles and strength to be top hooligan and later he was stabbed to death by another gangster.

I was thinking that the mirror was a good thing to reflect the image of stupidity, hypocrisy, immorality, and craftiness. I never heard any company leaders tell us what to do or bring new technology to us for advancement at all; just the reverse, as they tried to brainstorm us to tell them what and how to do the work in nuclear services.

I always thought that working in the nuclear services was much easier than taking an M16 rifle to shoot the Communist Việt Cộng in the jungle. I worked so hard to build the company from the substations to the turbine-generator and nuclear reactor control systems, etc....all by my own efforts.

I always worked at home with my own updated technology books. Whenever I came into my office, my jobs were done and I only had to take the time to put my works on paper. I always took care to incorporate the nuclear safety codes and design standards and never failed to make my designs state of the art with complete technical information and details that would well represent me in meetings with the site engineers to reduce phone calls or trips to the field.

One time my supervisor, Mr. L. M., told me to attend the video conference regarding reactor cooling systems at 1:00 P.M. I came into the meeting at exactly 1:00 P.M. Central Time, which was my time in Alabama. The guy who was the same rank as me, Mr. J. D., told me, "You are fired," in front of the video conference; I was totally surprised and did not say anything.

I later found out that it was supposed to be Eastern Time and that my supervisor, Mr. L. M., did not mention that to me. I kept waiting until the company meeting and told Mr. J. D., "I am not afraid to be fired, but you need to remember one thing—that the king of the jungle never scavenges."

Mr. J. D. flattered the project manager, Mr. G. M., excessively for putting me under a female supervisor, Mrs. J. G., and asked her to give me less than a one percent raise in wages that year in order to disgrace me. The sin of using a woman to disgrace a man is an unforgivable sin. Since the Stone Age, men always dominated and treated women as the last living creature on the food chain, but I look at women with spiritual respect as the tangible creators of the Universe and never let them fail by any means or any reason with my courage and my honest help. I worked very hard to make Mrs. J. G. look very good in management; she later was promoted to the Assistant of the Executive Vice President.

I was not left on my own, as the project manager, Mr. G. M., took me out from Mrs. J. G.'s group and put me under another manager, Mr. D. G., who gave me a new job (the replacement of the automatic frequency drive or AFD systems of the nuclear power plant hatch of the Georgia Power Co.), which cost twenty-one million dollars and ten thousand man-hours to finish. Later on, I told Mr. D. G. that I could manage to finish that project within eight thousand man-hours if I had good workers. Three weeks later my manager, Mr. D. G., called me in his office and told me, "Hue Le, I don't want you become anything else; Hue Le is just Hue Le."

I replied that I did not want to become anything else, either; I did fix all the wrong load calculations for many nuclear plants, which saved multi-millions of dollars for the company which would have had to replace expensive hardware caused by wrong calculations.

The largest nuclear power plant of Vogtle (a 13,050 mega-watt unit) was tripped for a couple of weeks and nobody, even those from General Electric and Westinghouse, could figure out what was wrong; it cost the company more than a million dollars per day.

The Beginning of Time

Technical Service Group called me in for troubleshooting; it took me several hours to figure out what was wrong. I looked at all the designs, which were correct, but the calibration of the extremely sensitive relay (affected per milli-volt), which controls the excitation of the advanced Generex system was set at the wrong level. This relay accurately controls the power output level of the turbine-generator and is associated directly with the nuclear reactor power output level. The problem with the power system was fixed and the unit was put back on line, normally within twenty-four hours.

My former project manager, Mr. T. A., came back to visit the nuclear plant E. I. Hatch project and told me, "Hue Le! You have been working five times more than anybody that has the same rank as you." However, I got paid less than any one of them.

Another manager, Mr. R. M., attended the same meeting with me regarding the sub-cooling nuclear fuel cracking problems; he blocked his ears when I suggested the redundant power supply to avoid loss of control power to maintain reliability for that system. After all, I had to protect my idea to build the backup control power to maintain reliability for the nuclear fuel systems. I was the engineer in charge of the main turbine-generator systems. I suggested that all the turbine-generator control systems be required to have an Uninterruptible Power Supply (UPS) to avoid tripping caused by an unreliable source of power; this system needed to be cool to prevent overheating. I tried to do both the designs of the UPS and the HVAC (Heat Ventilation and Air Conditioning) with system coordination for safe nuclear operation; the chief engineer of the Technical Support Department, Mr. D. S., took my design and made the power supply fuse larger than the well-coordinated value to show off his supreme knowledge, but it was totally wrong because of a short-circuit occurring downstream on the UPS that could send the fault current upstream and trip other system required for nuclear plant operation. I had to accept what I was told because I was a lower-ranking engineer.

I utilized the high technology to replace the old electro-mechanical control with a microprocessor control for the reactor building chiller systems; the design was already on paper and the system was implemented and operating in excellent condition, but none of the company engineers or the vendor's engineers (The Carrier Co.) could figure out and understand the new technology applied in the chiller systems. Whenever the managers or supervisors brought technical questions to me, they got the answers immediately without waiting a minute.

I carried out all the design works from electrical, instrument and control, nuclear system control, substations, turbine-generator control and the turbine building chiller systems, nuclear reactor systems, reactor building cooling systems, emergency diesel generators, etc., by myself and never asked for a raise or a promotion; however, unfair and jealous competition still gave birth to many ugly situations and immoral garbage that blocked my advancement.

The words being spoken about equal opportunity, fair and just, were just a sugarcoating that hypocrites always utilized to make an illusion for immoral cheating of the helpless people. I felt that I did not have room to place my feet in the company; I turned in papers for my retirement and wanted to go home to live as a red neck and forget all about material world. I raised a lot of vegetables and fruits from a variety of fruit trees and I sold them at some oriental food stores to make a living; I felt free and the best thing was being away from the trash of society.

One night God Cao Đài/Jehovah told me, "My great-grandson! You do not get angry and are not jealous. Thầy (your master) will give you everything that you have lost; the corruption will be punished by its own sin."

Nearly nine years after my retirement, all techniques, nuclear safety design methodology, and reliability control on nuclear support that I set forth and applied for nearly twenty-seven years—keeping our nuclear plants out of trouble—still carry heavy weight in nuclear services and support daily, at present time.

Sin and Punishment

Most of the old engineers of the company were lazy and never wanted to learn anything; they only wanted to become some sort of management and sit back to relax. I always tried to teach young people who just get out of college; they always wanted to learn and to do things. I wanted to have someone who could take my place after I left the company. For about twenty-six years, I provided commendable nuclear services to the company and no one among a couple thousand engineers of the company could compete with my work, even I had lost more than ten years as a freedom fighter in the jungle of Việt Nam, shooting the communist Việt Cộng.

The elitist engineering group started to play a new game, but they could not find any mistakes in my work in twenty-six years of services; they took the job that the young engineer in training from my group, Mr. J. M., and asked the senior engineer, Mr. N. S., to change it from right to wrong.

The engineer in training, Mr. J. M., brought the whole job and showed it to me. I told Mr. J. M. that his design was correct, but Mr. N. S. was the one checking his work and wouldn't sign off on it if he kept his original design and I had nothing to do with it. The elitist group then called me to attend a meeting; the high-ranking engineer, Mr. F. T., reprimanded me so badly as if he was the genius. I did not say a single word because by challenging the idiot, I would look like an idiot, as well; quietly, I asked Mr. J. M. to fix it as it was correct the first time.

Approximately seven years after my retirement, Mr. J. M. came to my house to pick up some fruit and he told me that Mr. N. S.'s wife left him, his house was burned to the ground, and then he got sick and passed away. I believe that any sin does have an equal punishment of its own.

I did provide two years of free service right after my retirement, even though they were unfair and unjust with respect to my excellent services and dedication.

(Please read a section of the comic named, "Dilbert," written by Scott Adams and published by the *Birmingham News* of Alabama, USA; the character that represents me is Wally, which is pronounced similar to my name Hue Le. [Wally is a very close pronunciation of my name, Hue Le, in Vietnamese.] I really do not know who Scott Adams is.)

I am pretty sure that nearly four years after I retired, the company collapsed, several CEOs, EVPs, VPs, and managers took turns leaving the company, and the Technical Support Department and Nuclear Design Department were falling apart and disappeared.

The nuclear services that I performed for the Bechtel Power Corporation nearly twenty-six years before now went back to them, and the company had to pay more than three times the amount for the same services that I provided for the company.

Some high-ranking officers and the contract company, Paragon, tried to get me back in the nuclear business. It would probably take another fifteen years to rebuild the company; I was nearly seventy years old and did not have that much time to do so. The company hired a lot of new engineers with high salaries; however, no work was done for more than three years.

Fairness and justice plus cause and effect came to play; my phone was ringing day and night and weekends for more than seven months. I disconnected my telephone and locked the gate of my house, cutting off all connections with the outside world. The high-ranking manager, Mr. B. J. G., desperately wanted to get me back on nuclear service and put things together for him to become the new CEO; he lost hope and finally he took his own life.

I promised myself that I would never be a slave for anybody on Earth, but for one and only God Cao Đài/Jehovah. Three years after my retirement, my house and a small piece of land used for a garden that were worth almost one hundred thousand dollars, now were worth more than two million dollars.

The intelligent people always think that they can utilize others' knowledge, labor, time, and fortune to build their own riches. They try very hard to absorb experiences and bear the unbearable in the pursuit of happiness and material things. Normally, they cheat or pillage other helpless people; however, they suffer a lot of shame and sorrow through punishment every day; all of these ugly and immoral activities just bring to them their goal of pursuing the unfair and unjust material life. Their lives could be compared to a little warehouse; if the warehouse exists they will exist, if the warehouse is gone they will be gone, as well.

The genius is totally different from the intelligent; the wisdom of the genius cannot be obtained from your parents, from your professors, from your teachers, from your friends, or from your wife.

To be a good offspring of God Cao Đài/Jehovah is the primary request for awarding the wisdom. Whenever God Cao Đài/Jehovah bestows the wisdom of genius on someone, He never takes it back and it is his or hers forever and He wants to see him or her with that award after life. It is not too hard, but not so easy to request the wisdom of genius from God Cao Đài/Jehovah; every one of us must be an honest human being, and share a common practice of charity, philanthropy, fairness, and justice for the whole world of human beings. This is also the key function to bring peace and love to the whole society of mankind. Faith in God Cao Đài/Jehovah is an essential source to communicate with Him and ask Him to forgive your sin and bestow the wisdom. If he or she does not get it (the wisdom of genius) today, tomorrow, or next year, they keep praying and carrying out the will of God Cao Đài/Jehovah (practicing charity, philanthropy, fairness, and justice) until he or she is qualified.

Whenever he or she is qualified to receive that spiritual award of wisdom, he or she will never worry about material things anymore; he or she will do everything with ease and material things will automatically follow after bestowing spiritual gifts from God Cao Đài/Jehovah. All painful problems of daily life will disappear and joy will naturally come with blessings.

I will bring some typical examples to prove that faith in God Cao Đài/Jehovah truly affects our lives. I had been fighting numerous deathly battlefields for more than ten years off and on in the jungle of Viet Nam; I always believed that God Cao Đài/Jehovah always drove off all deadly dangers away from me and protected me at all time, during more than ten years when death was just around the corner.

Turning an Unlucky Case into Good Fortune

One night after the Tet Overrun in 1968, I was a security officer in charge of patrolling for the whole city of Saigon; two 122-millimeter rockets hit right into the Congressional Building of the Republic of Viet Nam located on the Avenue of Lê-Lợi in Saigon City. My good-sized troop included police, military police, infantry, a signal group, marines, special forces, and paratroopers; we were moving toward the Congressional Building and approximately one hundred yards from the incident. I gave the order to stop the whole troop from advancing and starting to spread along the main Avenue of Lê-Lợi for security. After nearly forty minutes, nothing happened where we were, but rockets were still hitting around the city of Saigon. All of my soldiers were safe.

The city of Saigon was under curfew, and looking around, I saw the city was worse than the desert; nothing was moving, but all were my soldiers. I felt that night was as long a night as many where I was in the same kind of deadly incident when I was in the First and the Fourth Corps of the Republic of Viet Nam. After checking all my soldiers, I gave the order to my troops to move toward the Congress Building to see and report the damage. It was fortunate that nobody was killed or injured in that building; however, many large fires took place everywhere around the city and a lot of people were killed and injured in many communities. I could not stop my tears from falling; sorrow and sadness took over my heart and my soul and were embedded there forever.

After receiving the promotion of First Lieutenant, I was assigned to a new function as the Chief Officer of the Technological Research and Development Branch of the Signal Corps of the Army of the Republic of Viet Nam (ARVN). With the technical background as an assistant electrical engineer before being called for the draft, I was in charge of the calibration center, plus research and study of most of the technical problems for the Signal Corps of the ARVN.

During that time, a severe problem happened with our communication systems that all our PRC-10 utilized in most battlefields lost signal and friendly shootouts took place throughout four army corps. I was assigned to solve these problems; I started studying the overall design of the PRC-10 to figure out what was going wrong, and time was so constrained. I took many trips back and forth from the calibration center and the rebuild shop to take a close look at these problems. At last I found that the signal output from the amplifier was too low in amplitude—that caused the major problem in the communication systems.

I quickly focused on the design of that electronic component regarding parts, rebuilding technique, and the key function of gaining control. I asked the rebuild shop to throw away all the aging parts and damaged components and utilize good parts to rebuild, but no gain in control magnitude was to be found.

In order to meet the high demand of the PRC-10 for nationwide battlefields, I dedicated my time and concentrated on this job with all my effort to get it done and I found that the soldering heat was another factor to be considered. Finally, I used the heat compensation technique for the rebuild shop to follow and the problem was solved.

Then came the biggest honor that I had ever been awarded in my life; the top Commander of the Signal Corps of the ARVN, Colonel Phạm Văn Tiến, and the top American Signal Corps Commander Advisor in Viet Nam, Colonel Clair, came to the rebuild shop to testify my presentation of the problems.

After my illustration of the problems and my presentation of the technique for how to solve the problems, I was awarded a great party celebrated by both top signal corps commanders at the most famous restaurant in the city of Saigon, the Continental, and I was the one and only guest of honor. After all, Colonel Clair asked Colonel Phạm Văn Tiến to send me to the United States of America to study communication engineering at the Tennessee Technological University.

At the Tennessee Technological University, I finished two degrees in three and a quarter years; the BS degree was in Electrical Engineering and the Master of Science was in Plasma Physics. Communist North Viet Nam took over South Viet Nam in April 1975; during that time, I worked as a janitor for $1.25 per hour to support my education and I lived in a room 6 ft. x 10 ft. room without heater and air conditioning for $1.00 per day. I ate a couple of sandwiches and had a couple cups of coffee each day until I finished both the BS and MS in three years and three months.

I did have a fellowship for a PhD in Engineering Science, but I did not have time to finish it because my wife, Julie Phan Lê, and my son Phong Bằng Lê were left behind in Viet Nam under the cruel communist regime and wanted to get out of Viet Nam. The communist Viet Nam punished them pretty hard in retaliation for my cooperation with the Army of the USA; With their urgent demand, I had to leave my education to look for a job to rescue them as soon as possible.

After getting the job at the Southern Nuclear Operating Company, I worked through European governments and the USA government with the great help of the International Red Cross to get my wife and my son out of the hands of the cruel communist Viet Nam in 1979. Mr. Bob Mumpower, who recruited me to work for the Southern Nuclear Operating Company, later provided most of the paperwork to bring my family from Europe to the USA.

My wife, who was 38 years old, wanted to go to the School of Pharmacy at Samford University, and my son was 9 years old and had no education. The big problem was that both of them did not know English. I did try to borrow every penny that I could to support their education. I kept trying to refinance my home for its $47,500 value several times to support their education.

Finally with great effort my wife achieved her desire to become a professional pharmacist working at the Jefferson County Health Department in Alabama, and my son got his B.S. and M.S. Degree in Mechanical Engineering at the University of Alabama at Birmingham. He later took a special course of Computer Science from MIT and has worked for an air plane design company in California since then.

The Allocation of Wealth

There is no equipment that could measure accurately the skill, labor, knowledge, and dedication of anybody for a fair bargain; however, only honesty and moral principle could justify equitableness.

God Cao Dài/Jehovah gave us these social conventions and living principles:

1. Obey Nature
2. Respect Natural Prosperity Principles (Positive and Negative Law)
3. Practice Social Conventions of Morality, Honesty, Justness, Courteousness, Politeness, Sincerity, Fairness, and Balance.
4. Practice Charity, Philanthropy, Fairness, and Justice

If everyone in the whole world shared the simple practice of these principles, peace would come overnight and there would be love for everyone shortly thereafter.

The greedy character is the key factor that takes many people from being naturally born with moral principles to an unrighteous and atrocious way of life. If all the children were born in a peaceful and benevolent society, they would grow up as the most righteous characters wanted for all. Taking the same children just mentioned and putting them in a cruel community filled with pillagers and hooligans, they would grow up as pillagers or gangsters of the same ilk. Many times, children who grew up in an alcohol and drug-addicted family would become alcohol and drug addicted later, as well. Just in one generation, the largest population of children ever is alcoholic and drug addicted. They do not want to learn and to work; they just become a great burden of useless and sick people of society.

In martial art, one well-trained person would be equal to one hundred people who have not been trained; if one hundred people were well trained, their power would be equivalent to ten thousand people, as said in the ancient war time.

It will take more than four or five people to care for one sick person. If a large number of the children of a country are alcoholic and drug addicted, who will work to feed and take care of them? The strongest country will soon be the weakest nation because all families have to borrow whatever is needed to take care of them. This is the strongest power, able to deteriorate any country overnight. The creditors always have the upper hand in this situation, and it is an important thing for every one of us to remember.

It is very difficult to find a sick and lazy ant in any ant community; all of them are well trained to work to build their society. Ants of different kinds build their wonderful societies in peace and live with a high moral standard; they share hard work and equal reward to survive in good times and bad times. They protect their offspring better than human beings; they do not kill their offspring for honor, which is the first forbidden law of God Cao Dài/Jehovah. All children have the right to enjoy freedom, the right to exist, the right to live as a human being, and nobody can take away these rights from them.

The principles of peaceful coexistence:
- Freedom, fairness, and justness are the key sources of social equilibrium and motivation of creative things to make life easier and bring good things to society; they must be protected to conquer greed, selfishness, corruption, and immorality.
- All families must be strong and wealthy to build a strong country.
- Taking equal share to build the nation.
- Children are well educated and highly moral.
- Tax must be fair and enough for national defense and social needs and not for ill will.

- Government must always provide enough opportunities for all citizens.
- Providing law of equitableness with respect to moral principles.
- Fair and just practices must be exercised in all levels of society.
- Using charity and philanthropy to heal dolorous people.
- Helping the helpless people, but not creating scumbags to harm the country. As the saying goes: Giving people a fish is just providing for one meal; but teaching them how to fish, will benefit them for a lifetime. Giving them a fish and teaching them how to fish would be an ideal way of life.

Social equitableness is the most complex issue of the whole world:
- Some people were born in a poor family and never had a good opportunity to grow up as an educated person with good skills.
- Some people lost their parents when they were young, and poverty remains ahead as their future. A small number of them are successful and the rest are living poor.
- Some people were born in a rich family and had all kinds of opportunities to grow up as well-educated people with strong skills of moneymaking.
- Some people were born with plenty of opportunities, but never took them to build their future.
- Some people were born with inborn talents and skills; they did not have much education, but they could provide good things toward building our society.

There are multi-millions of jobs and every one of them requires different skill and labor to perform to reach an acceptable level. The technique of completion levels and the quality levels are more complex to classify; all of us attempt to figure out and write the law of equitableness to satisfy everyone in society. This is the most difficult thing to do and it has never been done since the Stone Age until now. Many laws have been written for this complex problem and a lot of them had loopholes which created unfair and unjust circumstances without correction.

The Beginning of Time

This complexity gave birth to socialists, communists, and labor unions to challenge capitalism, free-enterprise, and the free world. How far along the road of fairness and justness have people reached and how much equitability was achieved? The correct answer to that question is, none.

People in the world do not have enough food to survive, run around without clothes on, and wander without a shelter to be called home. In some places people still kill and eat their own races; killing takes place around the globe daily and the righteous people are outnumbered by the evil. Cheating and immoral practices give birth to violent chaos and wars in the whole world with no peace in sight. The road to peaceful coexistence unimaginable and may never be realized for many generations to come.

The key question is how the whole world could come together and build a peaceful coexistence in which freedom is a common practice and equitableness will bring a fair share to everyone who works for and deserves it. There are a million books written about a fair and just society, but none of them give us a way to achieve peaceful coexistence.

Working and Nature are two essential ways to support life. Working will bring good things to life, such as:

- Learning right from wrong to build better future necessities.
- Living with dignity and freedom.
- Self-development or inventing new technology for advancement to make life easier, not only for oneself, but for others, as well.
- A better chance to build strong and wealthy families and, in return, build a strong country.
- The combination of ideas of every individual's know-how in technology created from working will take any country into superior advancement.
- Knowing that doing nothing will bring nothing and sooner or later collapse will create ruin; a feeble individual will bring a feeble family and a feeble country, as well.
- Family, which is a building block of the country, and needs to be strong and wealthy.

- Busy work will keep at bay ill will, which is the source of dangerous crimes.
- Helping others.
- Producing, not begging for a living.

As the saying goes, family is the building block of a nation... lazy and sick families only weaken the country. Healthy and wealthy families build a strong country and a strong prosperity. If a leader of a nation tries to make people live in equal poverty, he or she directly destroys or weakens the building blocks of the nation; and, eventually, that country will collapse, as well. Strong blocks always enforce strong building. Healthy and wealthy people always build a healthy and wealthy country; the combination of all individual knowledge, experience, and skill is the primary source of a strong society structure and advancement. Stupid and lazy citizens only pull a country into collapsing; good woods always provide good buildings.

All human beings are born with a naturally-embedded benevolence and philanthropic character; however, they will pick up the good and the bad things from their surroundings as they grow up. The hardship of material life will create greed and selfishness that takes them far away from the moral and charitable way of life; as the world becomes more crowded, the scarcity of food supply and daily necessities will give birth to violent competition or killing and pillaging.

Our hands show the different level and ranking in society. When we hold something for work, for example, holding a shovel, the thumbs will always take fifty percent of the work force, and the rest of the work will be divided among the index fingers, the middle fingers, the ring fingers, and the little fingers.

The index fingers represent the group of people who have some political power, are well educated, do little work, and are very wealthy; they earn approximately twelve percent of wealth.

The middle fingers represent the highest ranking group in society, such as the kings or top leaders in the world of governance or of business, they are exceptionally wealthy, and very politically powerful; this group takes about eighteen percent of wealth.

The ring fingers represent the entourage group of the highest ranking of governance or business in society. Normally they are relative people

The Beginning of Time

of the top leaders, they have great political power and wealth; this group of people takes about thirteen percent of the wealth.

The little fingers represent the group of people who are very intelligent, but do little work, live simple, and do not want to be in political power. Normally they are genius and the real children of God Cao Dai/Jehovah; this group of people takes nearly seven percent of wealth.

The national treasure owns approximately forty-eight percent of the total property of the nation.

The thumbs represent fifty percent of the work force group, daily swapping their labor for a living; they share approximately two percent of the national property. If you keep your hands straight, you will see that the tips of the thumbs are just a little bit above the feet of other fingers, which shows them earning small shares. In the communist society, their leaders try to keep everyone equal; in order to make the unequal to be equal is like putting their finger's tips and the tips of their thumbs close to each other on a flat table. Everyone will see the tips of the thumbs and all of the other fingers will be equal, but the index fingers, the middle fingers, and ring fingers have to bend down in order to be equal. The thumbs and the little fingers do not need to bend to be equal because they are already the lowest part of society. The question is how long can the top leaders of the communist society keep bending down to be equal to their poor communist comrades?

If anyone can hold their hands with the tips of their thumbs and all of their fingers flatly equal for one day, then the top communist leaders could bend their backs to be equal to their comrades for one year, just for propaganda of cheating equality. In the communist society, the president of the communist party is exceptionally wealthy and has an absolute power to take anybody's life as he or she wishes. Now and then the communist leaders will pick up several laborers and promote them as heroes of labor, give them couple good meals, and after that sugarcoating promotions; these heroes will soon return to the simple labor slaves, no more or less. No matter how hard they work, laborers in a communist society sometimes are just not able to get enough food from hand to mouth. Immorality, lying, and cheating are common practices; they are atheistic and very cruel. The communists, satanic, and demonic worshipers will be punished equally for their sins. Sooner or later law of Nature will take all men back to their freedom, which is the best gift and always available from God Cao Đài/Jehovah.

Once again, after an hour or two of keeping your hands with all the tips flatly equal, every one of us will let our hands return to normal as naturally born—free and relaxed—and nobody can keep their hands bent forever.

There is a saying, "…obeying Nature will exist, against Nature will be destroyed or nonexistent…." And another saying, "All men are born free." Freedom is a silent word in the communist society.

In a free society, at least all of the human beings have the right to do things; to work and freely determine their own future, which is the unparalleled and unmatched treasure of all men. If we take a close look at the way of living in the whole world that the socialist, the communist, labor unions, and capitalist societies never create any laws that allocate wealth fairly and equitably; greed and selfishness are always searching for loopholes to interpret them to match ill will, which is the primary sources of wars. In general, the skill, knowledge, working level, ambition, free way of living or desire, and education or training, are the main complexities that create differences and biases.

Material is a temporary artificial thing, which does not belong to anybody forever and it will disappear as our first body (flesh body) is decaying. The good ideas of intelligent and the manual labor do have different values and it is not easy to clarify them for awarding. One good idea could make the whole world change. Normally a new good idea can bring good things to life and it deserves a good reward; but people cannot live without help from labor. What kind of measurement tool should be utilized to establish the scale for measuring value and its related reward in a fair and just manner?

These mutual facts are relatively bonded together and life cannot keep going without them. Freedom and the necessity of virtue could bring a fair, balanced, and peaceful society. Please take honesty and moral principles to create laws of fairness, justice, and equitableness for all and build the world in peaceful coexistence. God Cao Đài/Jehovah always wants His children to practice charity, philanthropy, fairness, and justice; and they are the real building blocks for eternal lives.

The Beginning of Time

A good thing to remember: Sin is not different from a jail, which confines the human soul for a long time in waiting for justice exercised by the law of Karma.

Building a good life for society and for your family is a good thing to do for every one of us; healthy families build strong country.

Please take charity, philanthropy, fairness, and justice as the spiritual medicine to cure the suffering world and build the society in peaceful coexistence; peace and beatitude will be at hand. The road to eternal life will open widely to welcome all of us.

Thank you for reading,
                Julie Phan Lê (Cao Lý Đạo)
                Professional Pharmacist
                Huệ Văn Lê (Cao Trí Đạo)
                Nuclear Engineer

The Law of Creation

Twin lotus flowers on one stem are extremely rare and could not be found anywhere in the whole world. These two lotus flowers represent the Law of Creation, which means that everything in the Universe must obey the Positive and Negative Principle to prosper and exist; otherwise all things will be destroyed or nonexistent. As observing everything around us daily, we will see that trees require pollination to prosper, and animals and human beings require the male and female incorporation to multiply and flourish, otherwise the whole world will be destroyed or the next generation will never come. All our equipment being used daily such as coffee machines, cars, radios, televisions, street light control systems, airplanes, space shuttles, air controller systems, etc…require electricity to operate; however, the electricity requires both positive and negative sources to operate and it is obvious that a single source of electricity (one positive source or one negative source alone) will not be able to do anything at all. This example shows that the opposition of the law of Nature (the Positive and Negative Principle) will self-terminate its own course of existence.

God Cao Đài/Jehovah awarded my family the spiritual gift of twin lotus flowers; we have twin grandchildren, a boy Casey, and a girl Kyra. This spiritual gift only came once and never again. Obeying Nature will allow us to flourish and prosper; going against Nature will cause us to self-terminate the next generation.

www.ingramcontent.com/pod-product-compliance
Lightning Source LLC
Chambersburg PA
CBHW061513180526
45171CB00001B/158